Reference Materials in Measurement and Technology

Sergei V. Medvedevskikh · Olga N. Kremleva ·
Irina E. Vasil'eva · Egor P. Sobina
Editors

Reference Materials in Measurement and Technology

Proceedings of the Third International Scientific Conference

Springer

Editors
Sergei V. Medvedevskikh
Ural Scientific Research
Institute for Metrology
Ekaterinburg, Russia

Olga N. Kremleva
Ural Scientific Research
Institute for Metrology
Ekaterinburg, Russia

Irina E. Vasil'eva
A.P. Vinogradov Institute of Geochemistry
Siberian Branch, Russian Academy
of Sciences
Irkutsk, Russia

Egor P. Sobina
Ural Scientific Research
Institute for Metrology
Ekaterinburg, Russia

ISBN 978-3-030-32533-6 ISBN 978-3-030-32534-3 (eBook)
https://doi.org/10.1007/978-3-030-32534-3

This Springer imprint is published by the registered company Springer Nature Switzerland AG
The registered company address is: Gewerbestrasse 11, 6330 Cham, Switzerland

Preface

This Springer collection features selected fundamental and applied research papers devoted to various issues involved with the development and application of reference materials and methods. These papers were presented at the Third International Scientific Conference *Reference Materials in Measurement and Technology*, which was held on 11–14 September 2018 in Ekaterinburg (Russia) under the auspices of the Federal Agency for Technical Regulation and Metrology (*Rosstandart*) and the Ural Scientific Research Institute for Metrology (*UNIIM*), a scientific methodological centre of the State service of reference materials for the composition and properties of substances and materials.

The conference was aimed at facilitating the exchange of scientific information between its participants about fundamental and applied research in the field of reference materials, as well as at attracting attention to reference materials as an important technical and regulatory basis for ensuring the uniformity and accuracy of measurements. In the context of knowledge economy, when digitisation comes to embrace not only industrial and social sectors, but also the entire quality control infrastructure, the maintenance of international uniformity of measurements becomes essential. In turn, the system aimed at providing uniformity of measurements also undergoes massive transformations concerned with increased informatisation and digitisation.

Under these circumstances, metrological activity is acquiring an infrastructural character by bolstering the national economy in all areas of activity, including health care, labour safety, pharmaceutical production, environmental protection, development and maintenance of diagnostic equipment. There is a clear trend towards the use of measuring instruments with distributed and virtual components, as well as cloud computing services and databases.

Intelligent systems are intended to simplify and optimise production and support the development of new technologies and substances. Already, new technologies permit more accurate data on measurement objects to be obtained, while advanced data processing methods offer improved approaches to analysing such measurement data. Experimental work devoted to the development of new reference materials is also becoming more accurate and efficient. As a result, the problem of both expanding the range and increasing the quality of modern reference materials is moving to the forefront. Another urgent task is the optimisation of mechanisms for forecasting the needs of states and society in terms of metrological measurement support, which can be achieved by increasing the competitiveness of domestic reference materials in international markets. Therefore, in the context of global digitalisation, key objectives of metrological support include not only monitoring but also a prompt response to the growing demands of national economies.

Leading experts in the field of measurements and reference materials from different countries are working on solving the aforementioned issues. The Third International Scientific Conference *Reference Materials in Measurement and Technology* brought together scientists, lecturers, postgraduate researchers and practitioners from various universities, research institutes and industrial companies. The conference was also attended by the representatives of ISO/REMCO, CITAC, COOMET, COMAR Central Secretariat and experts in the area of reference materials from Belarus, Germany, Israel, Kazakhstan, the Netherlands, Poland, Russia, South Africa, Switzerland and the USA.

The conference hosted meetings of 11 themed sections. During plenary, themed and poster conference sections, participants made over 80 presentations.

This collection contains 21 conference papers, which were selected during a rigorous double-blind peer-review process organised by the editors of this event. The papers presented in this collection were published in Russian in the peer-reviewed scientific and technical journal *Reference Materials*.

Ekaterinburg, Russia Sergei V. Medvedevskikh
Ekaterinburg, Russia Olga N. Kremleva
Irkutsk, Russia Irina E. Vasil'eva
Ekaterinburg, Russia Egor P. Sobina

Acknowledgements On behalf of the Organising Committee of the Third International Scientific Conference *Reference Materials in Measurement and Technology*, we express our sincere gratitude to all invited speakers, section chairpersons and conference participants for their valuable contribution to this event. We are also thankful to Anna Kveglis and Dr. Natalia Popova, Head of the Foreign Languages Department of the Institute of Philosophy and Law, Ural Branch of the Russian Academy of Sciences (Ekaterinburg) for their assistance in translating and preparing this book for publication.

We especially appreciate the work done by the members of the Conference Organising Committee I. E. Vasil'eva, A. I. Krylov, M. V. Okrepilov, N. P. Muravskaya, E. V. Kulyabina, who greatly helped in reviewing and editing these conference proceedings.

Contents

Development and Creation of Reference Materials

Development of Interlaboratory Comparison Test Samples and Reference Materials for the Quality Assurance of Air Composition Measurements

Anna M. Bogacheva, Olga B. Ponomareva and Yuliya V. Kanaeva

Abstract This article presents an approach to the development of interlaboratory comparison test samples and respective reference materials for use in the metrological assurance of measurements of hazardous substances in the ambient air. Solid and liquid sorbents, as well as air filters, were used in experiments. Measurement methods comprised photometry, atomic absorption spectrometry, inductively coupled plasma atomic emission spectrometry. An optimized algorithm is presented for developing multi-purpose test samples that can be used for the metrological assurance of a large number of methods for analysing ambient air quality. The described approach has been used to develop test samples for laboratory proficiency testing and respective reference materials, which can be used in validation of measurement techniques for analysing ambient air quality, calibration of measurement instruments, as well as in the internal and external quality control of results obtained during ambient air quality measurements.

Keywords Interlaboratory comparison (ILC) · ILC provider · ILC samples · Air media · Atmospheric air · Workplace air · Industrial emissions to the atmosphere · Laboratory test sample · Reference materials

Abbreviations

ILC Interlaboratory comparison
RM Reference materials
SPS State primary standard

A. M. Bogacheva (✉) · O. B. Ponomareva · Y. V. Kanaeva
Ural Scientifc Research Institute for Metrology (UNIIM), Yekaterinburg, Russia
e-mail: metod224@uniim.ru

S. Medvedevskikh et al. (eds.), *Reference Materials in Measurement and Technology*,
https://doi.org/10.1007/978-3-030-32534-3_1

1 Introduction

In Russia, interlaboratory comparisons (ILCs) for the purpose of laboratory proficiency testing are carried out by accredited ILC providers according to GOST ISO/IEC 17043-2013, GOST R ISO 13528-2010 and RMG 103-2010 [1–3]. ILCs aimed at monitoring the accuracy of experiments undertaken to measure the content of hazardous substances in the air, including atmospheric air, workplace air and industrial emissions to the atmosphere (Fig. 1), were first conducted by UNIIM as an ILC provider in 2015. Since then, ILCs of air quality measurements have been gathering a large number of participants (testing laboratories), which is an evidence of a great demand for the external quality control of air composition measurements. Therefore, the development of ILC samples (test samples for proficiency testing and control samples, hereinafter refereed to as test samples) seems to be a prospective research direction. It should be noted that such comparisons are also popular in other countries, including Canada, Brazil, Germany, Austria and the United Kingdom [4].

2 Results

Measurement techniques used for the analysis of the ambient air are based on various measurement methods, such as gas chromatography, spectrophotometry, atomic absorption spectrometry, inductively coupled plasma atomic emission spectrometry, etc. An important feature of analytical methods in the field of air quality control is the use of various so-called laboratory test samples, which are to be delivered to the laboratory.

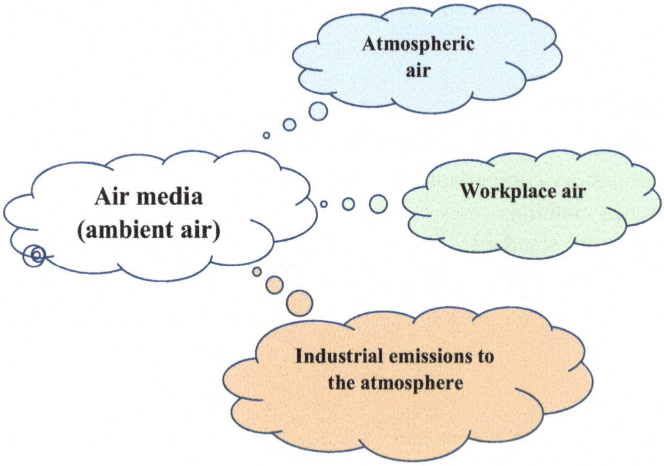

Fig. 1 Air media

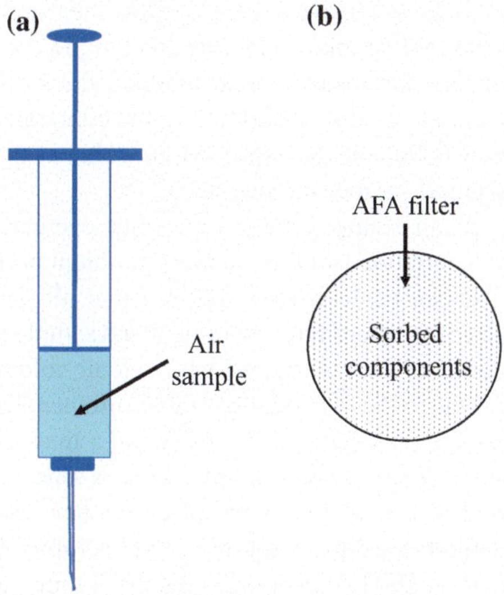

Fig. 2 Laboratory test samples: **a** for gas chromatography; **b** for spectrophotometry, atomic absorption spectrometry, inductively coupled plasma atomic emission spectrometry

A laboratory test sample can be presented in the form of:

(a) a collected air volume for performing gas chromatography. Figure 2a schematically represents such a sample placed in a syringe;
(b) components sorbed from the tested air for performing atomic absorption spectrometry with a flame or electrothermal atomiser, inductively coupled plasma atomic emission spectrometry or flame emission spectroscopy. These also include components analysed using a solid sorbent, an AFA filter or a liquid absorber. Figure 2b displays a schematic representation of such a sample with the measured components adsorbed onto an AFA filter.

In the present article, we consider a situation when a laboratory test sample is not a sample of the analysed object per se as shown in Fig. 2a, but rather a sample shown in Fig. 2b. Creating laboratory sample simulators is not an easy task, since it is not always clear what is being analysed. Thus, for ILCs, test samples simulating such laboratory samples are usually fabricated. In particular, components contained in the ambient air to be analysed are placed on a solid sorbent, a filter or in a liquid absorber. A similar approach is used when creating ILC samples in other countries. In Austria, for example, absorption solutions are used, while cellulose nitrate filters are applied in Germany.

Specific requirements are imposed on ILC test samples: they must be homogeneous and stable over the entire period of laboratory proficiency testing. In addition, UNIIM as an ILC provider requires a good test sample to be also versatile, i.e. being applicable in various analytical techniques (methods) used for the same component in the ambient air.

When participating in ILCs, laboratories use different techniques for determining the same component in the ambient air. It should be noted that every ILC provider performs extensive work to review measurement techniques used in various laboratories. When analysing these techniques, special attention is paid to the following aspects: the measurement method, the method of preparing samples for analysis, materials used to absorb the tested components, the measurement range of the analysed pollutant, the aspirated air volume, the amount of the laboratory test sample required for measurements.

As an example, Table 1 presents several techniques used to measure the iron mass concentration (total iron) in the ambient air [5–7].

It can be seen from Table 1 that all these measurement approaches [5–7] are based on different measurement and sample preparation methods, as well as different measurement ranges and aspirated air volumes.

Due to the multiplicity of measurement techniques, ILC providers aim to develop test samples that can be applied in a maximal number of measurement techniques. When creating a test sample, various aspects should be taken into account: materials used to absorb the analysed component (solid sorbents, filters, liquid absorbers); methods used for preparing samples for analysis (dry or wet ashing, etc.); measurement methods (spectrophotometry, atomic absorption, etc.), as well as the amount of the laboratory test sample required for particular measurements.

In order for the developed samples to be applicable in a maximal number of measurement techniques, the range of the analyte content in a laboratory test sample is determined for each technique. This is achieved taking into account parameters regulated by a particular measurement procedure, such as the aspirated air volume and the measurement range of the mass concentration of the analyte in the air. Afterwards, a measurement range common for all the techniques used to measure this pollutant is established. As a result, the value of the analyte content in a test sample should belong to this common range (Fig. 3).

In some cases, however, a common range for all the methods used cannot be defined (Fig. 4). Then several ranges are defined, for each of which a separate ILC sample is created.

Having analysed the techniques listed in Table 1, we determined that the total iron content in a laboratory test sample ranged from 0.3 to 1 mg. Following the abovementioned approach, a multi-purpose test sample with the analyte content varying within the established range can be created for use in all the techniques. To date, the UNIIM ILC provider has conducted 4 ILC rounds of determination of iron content (total iron) in the air. One ILC sample was created for each of the two rounds (as in Fig. 3), and two ILC samples with a different analyte concentration were created for each of the two more rounds (as in Fig. 4).

Figure 5 shows a test sample used in the 2018 ILC round for measuring the iron content in atmospheric air, workplace air and industrial emissions to the atmosphere for 12 measurement techniques in 38 laboratories. The versatility of the test sample was confirmed by the results of the ILC round, in which almost all laboratories obtained satisfactory results (Fig. 6).

UNIIM as an ILC provider develops similar samples for the external quality control of measurements carried out to determine the content of such air pollutants as manganese, dust, sulfuric acid, hydrochloric acid, nitrogen dioxide, ammonia, etc.

Table 1 Results of a random analysis of several measurement techniques used to monitor the iron mass concentration in the ambient air

No.	Measurement technique	Measurement method	Method of sample preparation for analysis	Sorbent	Measurement range of iron mass concentration in the monitored air (mg/m^3)	Aspirated air volume (dm^3)
1	**MVI 07-04** Measurement procedure for determining the iron mass concentration (Fe^{+3}) in industrial emissions by the photometric method	Photometry	Wet ashing	AFA filter	From 1.0 to 1500 incl.	300
2	**MU 4945-88** Guidelines for the determination of hazardous substances in the welding aerosol	**p. 3.1.** Spectrophotometry	Dry ashing	AFA filter	From 1.0 to 15 incl.	200
		p. 3.4. Atomic Absorption Spectrometry	Wet ashing		From 0.01 to 10 incl.	From 100 to 300 incl.
3	**PND F 13.1.66-09** Measurement procedure for determining the elemental mass concentrations in industrial emissions by inductively coupled plasma atomic emission spectrometry	Inductively coupled plasma atomic emission spectrometry	Wet ashing	AFA filter	From 0.00125 to 25 incl.	From 200 to 600 incl.

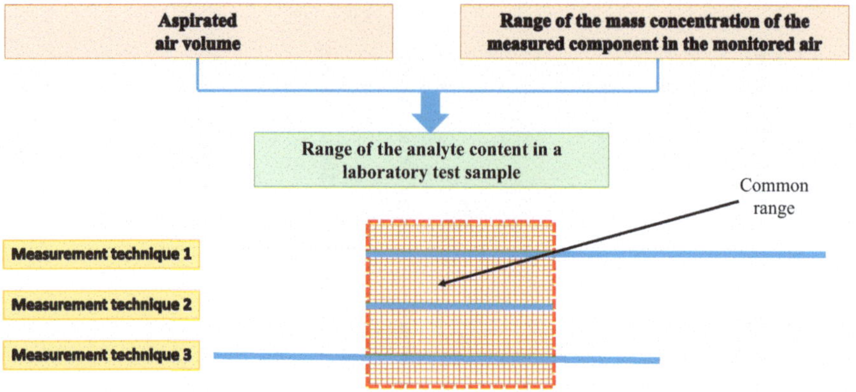

Fig. 3 A scheme for determining a measurement range common for all the measurement techniques used

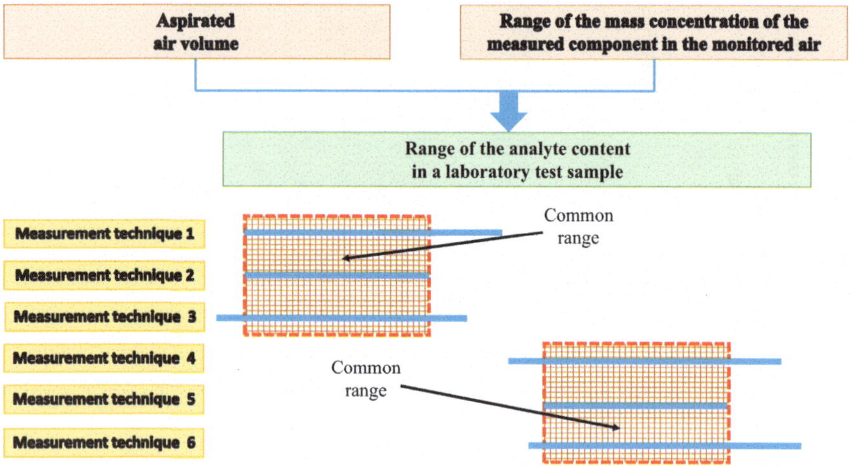

Fig. 4 A scheme for determining a measurement range common for all the measurement techniques used

For every specific ILC, a test sample with the specified analyte content is created, taking into account the measurement techniques used by participant laboratories (as is shown in Figs. 3 and 4).

The assigned values of such ILC test samples can be established on the basis of the measurement results obtained by the ILC participants, indicating the metrological comparability of measurement results obtained by ILC participants (according to the International Vocabulary of Metrology, term 2.46) [8]. At the same time, the assigned values of ILC test samples can be obtained using the methods of their evaluation ensuring their metrological traceability (term 2.41) [8] to the standards of measurement units (for example, SS 176-2013—SPS of the units of mass (molar)

Fig. 5 A test sample for measuring iron mass concentration in the ambient air

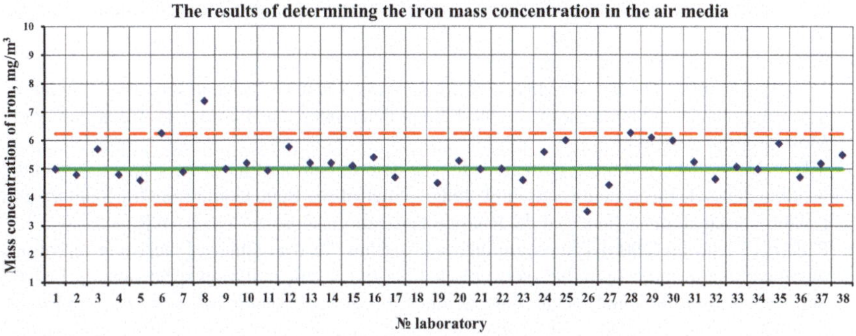

Fig. 6 The results of determining the iron mass concentration in the air

fraction and mass (molar) concentration of the component in liquid and solid substances and materials based on coulometry)[1] [9], which allows these samples to be used for the metrological assurance of measurements and the development of similar RMs.

---

[1]SPS 176-2013 State primary standard of the units of mass (molar) fraction and mass (molar) concentration of a component in liquid and solid substances and materials based on coulometry//Ural Scientific Research Institute for Metrology (UNIIM) [site]. https://uniim.ru/gauge_fgup_uniim/.

3 Conclusion

RMs developed using the described approach can be applied for the validation of measurement techniques for analysing ambient air quality, calibration of measurement instruments and the internal and external quality control of measurements undertaken to monitor the ambient air quality [10]. The use of such samples increases the metrological assurance of air quality measurements, which eventually contributes to the reliability of information about the current state of the environment.

Acknowledgements The authors express their sincere gratitude to the anonymous reviewers and the Editor for their highly professional and benevolent attitude shown in the preparation of our materials for publication.

References

1. GOST ISO/IEC 17043-2013 (2014) Conformity assessment. General requirements for the conduct of proficiency testing. Standartinform, Moscow, 33 p (in Russ.)
2. GOST R ISO 13528-2010 (2012) Statistical methods. Application for experimental verification of competence through interlaboratory comparative tests. Standartinform, Moscow, 54 p (in Russ.)
3. RMG 103-2010 (2011) Testing the qualifications of testing (measuring) laboratories that test substances, materials and environmental objects (in terms of composition and physicochemical properties) by means of interlaboratory comparative tests. Standartinform, Moscow, 38 p (in Russ.)
4. EPTIS database. http://www.eptis.org/
5. MVI 07-04 (2004) The measurement procedure of the iron mass concentration (Fe^{+3}) in industrial emissions by the photometric method. Saint Petersburg, The Ecological Research Center, Moscow, 12 p (in Russ.)
6. MU 4945-88 (1992) Method guidelines for the determination of hazardous substances in the welding aerosol. Information and Publishing Center of the State Committee for Sanitary and Epidemiological Surveillance of the Russian Federation, Moscow, 110 p (in Russ.)
7. PND F 13.1.66-09 (2009) The measurement procedure for the elemental mass concentrations in industrial emissions by inductively coupled plasma atomic emission spectrometry. Federal Service for Environmental, Technological and Nuclear Supervision, Moscow, 16 p (in Russ.)
8. ISO/IEC GUIDE 99: 2007 (2007) International vocabulary of metrology. Basic and general concepts and related terms (VIM), 108 p
9. GOST ISO Guide 35-2015 (2017) Reference materials. General and statistical principles of certification (validation). Standardinform, Moscow, 58 p (in Russ.)
10. Kropanev AY (2018) Methodologies for analysis of air environment. Certification and arrangement of intra laboratory control of the quality of analysis results. Industrial laboratory. Diagn. Mater. **84**(8):65–70. https://doi.org/10.26896/1028-6861-2018-84-8-65-70 (in Russ.)

Metrological Assurance of Dioxin Measurements. A Certified Reference Material for Measuring the Dioxin Mass Fraction in Animal Fat

Alexandra G. Budko, Alena Y. Mikheeva and Anatoliy I. Krylov

Abstract This article presents information on the development of a new type of certified reference material (CRM) of the mass fraction of dioxins in animal fat, assured with metrological traceability. A procedure for preparing and certifying the CRM is provided. GC-LRMS in combination with isotopic dilution was used as an analytical method for establishing certified characteristics. The uncertainties induced by the CRM inhomogeneity and instability were estimated in order to account for their contribution to the combined uncertainty of the obtained certified values. In addition, the results of interlaboratory comparisons (ILC) studies for an experimental batch of the developed CRM are presented. Measurement results provided by ILC participants are shown to be in good agreement with the certified characteristics of the CRM. The certified values of the mass fractions of dioxins (PCDF group) in the CRM are established to vary from 1 to 1000 ng/kg, with the relative expanded uncertainty ranging from 10 to 20%.

Keywords Certified reference materials · Dioxin · Certified value · Uncertainty · Traceability · Mass balance method · Inhomogeneity · Stability

Glossary of Abbreviations

HpCDD	Heptachlorodibenzo-p-dioxins
HpCDF	Heptachlorodibenzofurans
HxCDF	Hexachlorodibenzofurans
HxCDD	Hexachlorodibenzo-p-dioxins
OCDD	Octachlorodibenzo-p-dioxin
OCDF	Octachlorodibenzofuran
PeCDD	Pentachlorodibenzo-p-dioxins

A. G. Budko (✉) · A. Y. Mikheeva · A. I. Krylov
D.I. Mendeleyev Institute for Metrology (VNIIM), Saint Petersburg, Russia
e-mail: aa@b10.vniim.ru

A. I. Krylov
e-mail: akrylov@b10.vniim.ru

S. Medvedevskikh et al. (eds.), *Reference Materials in Measurement and Technology*,
https://doi.org/10.1007/978-3-030-32534-3_2

PeCDF Pentachlorodibenzofurans
TCDD Tetrachlorodibenzo-p-dioxins
TCDF Tetrachlorodibenzofurans

1 Introduction

The determination of the mass fraction of polychlorodibenzo-p-dioxins (PCDD) and polychlodibenzofurans (PCDF) in various environmental objects and food products presents undisputed scientific and practical interest. Since the middle of the 20th century, when the dioxin problem was first realised, considerable attention has been paid to dioxin measurement in various objects [1–7].

PCDD and PCDF have never been the target products of organic synthesis, emerging exclusively as impurity components during various technological processes (pesticide production, both chemical and household recycling, pulp bleaching in paper production, etc.). Secondary sources (deposits) of PCDD and PCDF comprise environmental objects, in which these substances have been stored for decades. These include, e.g., wood impregnated with pentachlorophenol, transformer oils containing polychlorinated biphenyls (PCB), active sludge or composts, waste dumping and garbage landfills, as well as contaminated bottom sediments. Such reservoirs can potentially promote the repeated emission of PCDD and PCDF to the environment [8].

PCDD and PCDF belong to the group of persistent organic pollutants (POPs), thus being subject to the Stockholm Convention [9]. This document prioritises the application of global environmental protection measures to the POPs, since these ecotoxicants pose a serious threat to human health and biota as a whole.

Congeners of the PCDD and PCDF groups (17 out of 210) with chlorine atoms in positions 2, 3, 7 and 8 of the base molecule (hereinafter referred to as dioxins) characterised by a pronounced biological activity are responsible for the long-term contamination of the biosphere, thus being recognised as toxic. Being accumulated in living organisms, dioxins are capable of modifying biochemical processes. These substances exhibit potent mutagenic, immunosuppressive, carcinogenic and embryotoxic properties, causing disruption of the nervous and cardiovascular systems, as well as of the gastrointestinal tract and liver. This is the reason for them to be classified as Class 1 hazards [10]. Such specific properties cause reasonably stringent standards for the mass fraction of dioxins in foodstuffs (see Table 1) both in Russia and in the European Union (EU).

Due to the high danger of dioxins and, accordingly, stringent standards (see Table 1), limits of their determination should be significantly lower than those typically applied to other ecotoxicants. Thus, regardless of the object of study, the measurement of dioxins is a complex analytical task, the solution of which requires the development and implementation of separation and detection methods that provide for an isomer-specific determination of ultra-low (picogram) quantities

Table 1 Hygienic requirements for foodstuffs [11, 12]

Foodstuffs	Maximum levels, pg/g fat	
	Russia	EU
Meat and meat products (bovine animals and sheep)	3	2.5
Meat and meat products (pork)	1	1
Hen eggs and egg products	3	2.5
Raw milk and dairy products	3	2.5
Marine oils	2	1.75
Vegetable oils and fats	0.75	0.75

of dioxins against the background of substances present in the matrix in significantly higher concentrations.

Conventional methods for measuring dioxins include extraction of analytes from the matrix using suitable extraction techniques, multi-stage purification of the extract on mineral sorbents, concentration of the extract and instrumental isotopic dilution analysis via gas chromatography low or high resolution mass spectrometry (GC-LRMS or GC-HRMS) or tandem mass spectrometry (GC-MS/MS). These algorithms form a basis for the development of state methods for measuring the mass fraction of dioxins in both Russia and the USA, Japan, the European Union, etc. Some techniques are provided in works [13–19].

Instrumental methods mentioned above involve the appliance of appropriate reference materials with certified concentrations of analytes for the construction of calibration characteristics, as well as for the validation and verification of measurement methods and control of the accuracy of measurement results. It is important to note that the use of measuring and validation instruments assured by metrological traceability with regard to the corresponding units of quantities not only guarantees the required accuracy and reliability of measurement results, but also ensures their comparability across various analytical laboratories and their recognition at the international level (Agreement CIPM MRA dated 10/14/1999 [20]).

In organic analysis, the traceability assurance is based on the application of pure organic substances with the certified mass fraction of the main component and corresponding uncertainty. From this perspective, mass balance method manifests itself as a traditional and internationally recognised indirect method for determining the purity of organic compounds. It involves the measurement of four possible groups of impurities (related compounds, water, volatile organic compounds and non-volatile substances) in pure organic substance and the subsequent calculation of the mass fraction of the main component according to the formula "100% minus the sum of impurities" [21, 22]. Thus, the first prerequisite for the formation of the traceability chain is the presence of pure organic matter in sufficient quantities.

As mentioned above, dioxins are not commercial products. Pure organic substances of dioxins are synthesised in limited quantities by several foreign laboratories exclusively for analytical use. At the moment, dioxin measurements can be performed

Table 2 Certified reference materials for dioxin measurements. Solutions for calibration of analytical equipment [23, 24]

CRM number	Description	Manufacturer
BCR-614	Polychlorodibenzo-p-dioxins and polychlorodibenzofurans in n-nonane (Nine solutions)	JRC-IRMM
CRM 10481-2014	PCDD/PCDF in nonane (DF-10)	STC RCSH Russia
CRM 10480-2014	Isotope labeled PCDD/PCDF in nonane (DF-9)	STC RCSH Russia
CRM 10479-2014	Isotope labeled PCDD/PCDF in nonane (DF-8)	STC RCSH Russia
CRM 10477-2014	Isotope labeled PCDD/PCDF in nonane (DF-6)	STC RCSH Russia
CRM 9634-2010	PCDD/PCDF in nonane (DF-5)	STC RCSH Russia
CRM 9633-2010	Isotope labeled PCDD in nonane (DF-4)	STC RCSH Russia
CRM 9632-2010	Isotope labeled PCDD/PCDF in nonane (DF-3)	STC RCSH Russia
CRM 9630-2010	Isotope labeled PCDD/PCDF in nonane (DF-1)	STC RCSH Russia

using the certified reference material (CRM) produced by the Joint Research Centre—Institute for Reference Materials and Measurements (JRC-IRMM, Belgium), as well as the Russian state reference materials of dioxin solutions developed and manufactured by the Scientific-Technical Centre of Radiation and Chemical Safety and Hygiene of the Federal Medical Biological Agency of Russia (STC RCSH). Table 2 provides information on the available CRMs of dioxin mass fraction.

Out of all the CRMs presented in Table 2, only BCR-614 solutions produced by JRC-IRMM are characterised by metrological traceability (purity of individual substances determined by the mass balance method, identity confirmed by the ^{1}H-NMR analysis).

Along with the CRMs listed in Table 2, non-traceable standard solutions (Standard Solutions) of dioxins produced by the Wellington Laboratories, Inc. (Canada) [25] and Cambridge Isotope Laboratories, Inc. (USA) [26] are commercially available, with the majority of analytical laboratories around the world using them for routine measurements. However, it should be noted that the mass fractions of dioxins in these solutions are not certified and can only be used as reference values.

Matrix CRMs with the certified dioxin mass fraction value suitable for validation and verification of measurement methods and control of their accuracy are numerous and comprise commercial products of JRC-IRMM and the National Institute of Standards and Technology (NIST, USA). Information on available matrix CRMs is presented in Table 3.

It can be seen from Table 3 that, among the existing CRMs for measuring the concentrations of dioxins, only (CRM) BCR-614 produced by JRC-IRMM is provided with metrological traceability to the corresponding units. The data presented in Table 3 indicate the lack of matrix CRMs with certified dioxin mass fraction values in Russia.

Considering the fact that dioxins in the form of pure substances are difficult to access and extremely expensive, in this work, we did not aim to certify dioxins as

Table 3 Matrix RM with certified dioxin mass fraction [23, 27]

CRM number	Description	Manufacturer	Traceability data
BCR-490	Fly ash	JRC-IRMM	BCR-614 (IRMM)
BCR-529	Industrial (sandy) soil	JRC-IRMM	BCR-614 (IRMM)
BCR-530	Industrial (clay) soil	JRC-IRMM	BCR-614 (IRMM)
BCR-607	Milk powder	JRC-IRMM	BCR-614 (IRMM)
BCR-615	Fly ash (low level)	JRC-IRMM	BCR-614 (IRMM)
BCR-677	Sewage sludge	JRC-IRMM	BCR-614 (IRMM)
SRM 1944	New York/New Jersey waterway sediment	NIST	No data*
SRM 1953	Organic contaminants in non-fortified human milk	NIST	No data*
SRM 1954	Organic contaminants in fortified human milk	NIST	No data*
SRM 1957	Organic contaminants in non-fortified human serum (freeze-dried)	NIST	No data*
SRM 1958	Organic contaminants in fortified human serum (freeze-dried)	NIST	No data*
SRM 2786	Fine atmospheric particulate matter (mean particle diameter < 4 μm)	NIST	No data*
SRM 2787	Fine atmospheric particulate matter (mean particle diameter < 10 μm)	NIST	No data*

*Dioxins in these materials are characterised by reference, rather than certified, values. These values partially meet the NIST criteria for certification, since not all sources of uncertainty may be taken into account in data processing [27]

pure substances. Our purpose was to develop a new type of compositional CRMs of the mass fraction of dioxins in animal fat with assured metrological traceability. CRM BCR-614 [23] produced by JRC-IRMM was used as a reference to ensure traceability of the relevant units of quantities with regard to the type of quantity (mass fraction or concentration of PCDD/PCDF). Pork fat was chosen as a matrix for the CRM under development, representing a proper model of fat-containing food products of animal origin.

2 Experimental Part

A CRM for measuring the concentration of dioxins in animal fat was prepared from meat and bone meal naturally contaminated with dioxins at the Research Department of State Standards in the Field of Organic and Inorganic analysis, D.I. Mendeleyev Institute for Metrology (VNIIM), Russia. The lipid fraction containing dioxins and extracted from meat and bone meal was added to the melted lard. The as-obtained liquid matrix was thoroughly homogenised and packaged in airtight glass vessels (vials). Each vessel contained 5 g of the liquid matrix. Samples for instrumental analysis were prepared using the generally accepted method described in works [13, 28]. The certified reference material BCR-614 produced by JRC-IRMM was used for the calibration of the analytical equipment and as an isotopically labelled analogue.

Measurements of dioxin mass fraction values were carried out by two analytical systems, GC-LRMS and GC-MS/MS, using the method of isotope dilution. It should be noted that mass spectrometry in combination with isotope dilution is widely recognised as the primary method [29–32], which ensures the best performance from the standpoint of measurement accuracy.

GC-LRMS measurements were carried out in the mode of registration of selected ions (SIM), while the GC-MS/MS analysis was performed in the mode of multiple reactions monitoring (MRM). Mass-spectrometric operating conditions are shown in Table 4. The parameters of the chromatographic separation of dioxins are presented in Table 5.

3 Results and Discussion

As noted above, meat and bone meal naturally contaminated with dioxins was used to prepare the CRM. As a result, the congener profile of dioxins in the CRM exhibited a number of specific features: the concentration of PCDF congeners ranged from 1 to almost 1000 ng/kg, while the concentration of the PCDD group varied within the same order not exceeding 4 ng/kg.

In order to establish the CRM certified values and to investigate the CRM stability and homogeneity, the following PCDF congeners present in the RM were selected considering the specificity of the congener profile: 2,3,7,8-TCDF, 2,3,4,7,8-PeCDF, 1,2,3,4,7,8-HxCDF, 1,2,3,6,7,8-HxCDF, 2,3,4,6,7,8-HxCDF, 1,2,3,4,6,7,8-HpCDF, 1,2,3,4,7,8,9-HpCDF. Measurements were performed using the GC-LRMS method.

The mass fraction of each dioxin congener was calculated according to the equation:

$$w_{an_i} = \frac{A_{an_i} \times m_{IS_i}}{A_{IS_i} \times RF_i \times m_s}, \tag{1}$$

where

w_{an_i} is the mass fraction of the i-th analyte (congener) in an analytical sample weight, ng/g;

A_{an_i} is the peak area of the i-th congener, c.u.;

A_{IS_i} is the peak area of the internal standard of the i-th congener, c.u.;

m_{IS_i} is the weight of the internal standard of the i-th congener, added before preparing the sample for analysis, ng;

m_s is the analytical sample weight, g;

Table 4 Operating conditions for mass-spectrometric measurement

Congeners	SIM		MRM		
	M1/M2 (m/z)	Natural ratio (identification parameter)	Precursor ion/Product ion (m/z)	Precursor ion/Product ion (m/z)	Collision cell energy (V)
Analyte group					
TCDD	320/322	0.77	320/257	322/259	24
TCDF	304/306	0.77	304/241	306/243	33
PeCDD	356/358	1.32	356/293	354/291	25
PeCDF	340/342	1.32	340/277	338/275	35
HxCDD	390/392	1.24	390327	392/329	25
HxCDF	374/376	1.24	374/311	376/313	35
HpCDD	424/426	1.05	424/361	426/363	25
HpCDF	408/410	1.05	410/347	408/345	36
OCDD	458/460	0.89	458/395	460/397	26
OCDF	442/444	0.89	442/379	444/381	35
Isotopically labelled group of analogues					
$^{13}C_{12}$-TCDD	332/334	0.77	332/268	334/270	24
$^{13}C_{12}$-TCDF	316/318	0.77	316/252	318/254	33
$^{13}C_{12}$-PeCDD	368/370	1.32	366/302	368/304	25
$^{13}C_{12}$-PeCDF	352/354	1.32	352/288	350/286	35
$^{13}C_{12}$-HxCDD	402/404	1.24	404/340	402/338	25
$^{13}C_{12}$-HxCDF	384/386	1.24	386/322	388/324	35
$^{13}C_{12}$-HpCDD	436/438	1.05	438/374	436/371	25
$^{13}C_{12}$-HpCDF	418/420	1.05	420/356	422/358	36
$^{13}C_{12}$-OCDD	470/472	0.89	470/406	472/408	26
$^{13}C_{12}$-OCDF	454/456	0.89	454/390	456/392	35

RF_i is the response factor of the *i*-th congener relative to the internal standard of the *i*-th congener.

The response factor RF_i was calculated as follows:

$$RF_i = \frac{A_{an_i} \times m_{IS_i}}{A_{IS_i} \times m_{an_i}}, \quad (2)$$

where

A_{an_i} is the peak area of the *i*-th congener in the calibration solution, c.u.;
A_{IS_i} is the peak area of the internal standard of the *i*-th congener in the calibration solution, c.u.;
m_{an_i} is the weight of the *i*-th congener in the calibration solution, ng;

Table 5 Operating conditions for chromatographic separation of dioxins

Parameter	Characteristic
Column	Rtx-Dioxin2, 60 m × 250 mkm × 0.25 mkm
Carrier gas	Helium
Flow (mL/min)	1
Injection volume (μL)	1
Inlet, mode	Splitless
Inlet temperature (°C)	280
Initial temperature of column (°C)	120
Hold time (min)	0
Rate (°C/мин)	20
Ramp 1 (°C)	200
Hold time (min)	1
Rate (°C/мин)	2
Ramp 2 (°C)	290
Hold time (min)	40
Ionisation method	Electron ionisation (EI)

m_{IS_i} is the weight of the internal standard of the *i*-th congener in the calibration solution, ng.

The calculation of the relative combined standard uncertainty for each congener was performed according to the formula:

$$\frac{u_w}{w} = \sqrt{\left(\frac{u_{m_{IS}}}{m_{IS}}\right)^2 + \left(\frac{u_{m_s}}{m_s}\right)^2 + \left(\frac{u_{RF}}{RF}\right)^2 + \left(\frac{u_{w_{av}}}{w_{av}}\right)^2}, \tag{3}$$

where

$\frac{u_{m_{IS}}}{m_{IS}}$ is the relative standard uncertainty due to the weight of the internal standard;
$\frac{u_{m_s}}{m_s}$ is the relative standard uncertainty due to sample weight;
$\frac{u_{RF}}{RF}$ is the relative standard uncertainty caused by the response factor determination;
$\frac{u_{w_{av}}}{w_{av}}$ is the relative standard uncertainty of the mean value (RMSD of measurement results).

The greatest contribution to the combined standard uncertainty is shown to be made by the uncertainty of the mean of mass fraction $\left(\frac{u_{w_{av}}}{w_{av}}\right)$ for the lower range of values and the uncertainty of determining the response factor $\left(\frac{u_{RF}}{RF}\right)$ for the rest of the value range. The relative combined standard uncertainty was calculated for each congener individually, ranging from 10 to 20% for the entire measurement range.

In order to determine the contribution of uncertainty induced by the CRM inhomogeneity (u_h), 10 out of 300 vials were randomly selected from the entire CRM

Table 6 Results of determining the mass fraction of 1,2,3,4,6,7,8-HpCDF in the study of the CRM homogeneity

No.	Mass fraction (%)					
	1	2	3	4	5	6
1	47.0	46.8	50.5	50.1	47.7	50.4
2	49.0	47.7	46.5	50.1	49.4	47.9
3	50.3	49.8	46.5	47.9	48.6	48.9
4	46.4	47.6	46.0	46.7	46.6	49.8
5	48.1	47.4	48.9	49.3	50.0	47.4
6	48.9	47.5	47.8	47.5	47.7	50.8
7	48.3	47.8	48.2	48.9	47.9	47.1
8	49.3	50.1	46.0	48.7	47.6	46.5
9	48.3	49.8	48.0	49.3	51.0	50.9
10	46.6	46.0	48.3	47.3	46.0	47.6

batch. Two samples of 2 g each were taken from each vial to perform three parallel instrumental measurements. The data obtained was processed using a one-way ANOVA test taking into account the provisions given in works [33, 34]. An example of a data set for the CRM homogeneity study is presented in Table 6. The results of the data analysis are presented in Table 7.

The worst of the obtained values equalling 0.36% was accepted to be the relative standard uncertainty due to the CRM inhomogeneity u_h. Thus, the inhomogeneity contribution to the combined uncertainty was deemed to be negligible.

The determination of the shelf life of the developed CRM, as well as the estimation of uncertainty caused by long-term stability (u_{stab}) of dioxins in the fat matrix, were carried out for each congener by the classical method of stability analysis according to the algorithm provided in the guidelines [33]. The residual shelf life was determined by the isochronous method. An example of the results of measuring the mass fraction of 2,3,4,6,7,8-HxCDF in stability experiments is presented in the form of a diagram in Fig. 1.

On the basis of the experimental data obtained, the greatest contribution from instability (u_{stab}) was found to be equal to 1.8%, which is negligible. Thus, CRM samples were considered stable under the following storage conditions: a dark place, ambient temperature of (20 ± 5) °C and no exposure to corrosive media. The shelf life of the CRM is found to be 10 years.

The mass fractions of seven congeners of the PCDF group were established as certified RM values. The values of the mass fractions of PCDD congeners and the total toxicity of a sample expressed in terms of international toxicity equivalent (TEQ) are given as reference. Characteristics of the CRM of the mass fraction of dioxins in animal fat are presented in Table 8.

An experimental batch of the CRM of the dioxin mass fraction in animal fat was used as an object of interlaboratory comparisons (ILC), which were supported

Table 7 Analysis of the data obtained in the study of the CRM homogeneity (for 1,2,3,4,6,7,8-HpCDF)

ANOVA									
No.	Count	Sum	Average	Variance	Median	Asymm.	S^2	SD	RSD
1	6	292.5100	48.7517	3.7500	48.8950	−0.085	1.4E−04	1.2E−02	8.4
2	6	290.6800	48.4467	3.6100	48.4500	−0.224	3.4E−04	1.8E−02	12.9
3	6	292.0800	48.6800	3.7700	48.7400	−0.539	5.7E−04	2.4E−02	16.7
4	6	283.0200	47.1700	3.7300	46.6400	1.786	4.0E−04	2.0E−02	14.8
5	6	291.1000	48.5167	2.6200	48.5100	0.233	5.7E−04	2.4E−02	14.7
6	6	290.1100	48.3517	3.3500	47.7350	1.693	9.7E−04	3.1E−02	18.2
7	6	288.2600	48.0433	1.8000	48.0750	−0.273	1.1E−03	3.4E−02	21.2
8	6	288.0500	48.0083	4.1300	48.1400	−0.088	4.5E−04	2.1E−02	12.1
9	6	297.2800	49.5467	3.0100	49.5750	−0.074	7.3E−05	8.5E−03	5.8
10	6	281.7100	46.9517	2.3300	46.9350	0.297	2.1E−04	1.5E−02	10.0
Source of variation				SS	df	MS	F	P-value	F crit
Between groups (SS_h)				31.15	9	3.46	2.05	0.053	2.07
Within groups (SS_e)				84.51	50	1.69			
Provided $F < F_{crit}$, inhomogeneity induced by standard deviation is negligible									
Total				115.66	59				

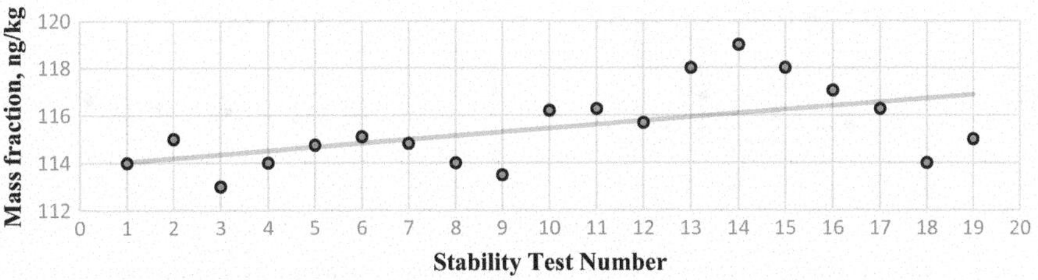

Fig. 1 A diagram of data obtained during the long-term stability study of the CRM (for 2,3,4,6,7,8-HxCDF). The measurements were carried out on a quarterly basis, starting with the II quarter of 2013 (Test no. 1—II quarter of 2013) [16]

Table 8 Characteristics of the CRM of the mass fraction of dioxins in animal fat

Congener	Mass fraction (ng/kg)	Expanded uncertainty (k = 2) (ng/kg)
Certified values		
2,3,7,8-TCDF	1.18	0.24
2,3,4,7,8-PeCDF	823	76
1,2,3,4,7,8-HxCDF	215	20
1,2,3,6,7,8-HxCDF	116	13
2,3,4,6,7,8-HxCDF	115	14
1,2,3,4,6,7,8-HpCDF	47.0	4.6
1,2,3,4,7,8,9-HpCDF	9.27	1.06
Reference values		
2,3,7,8-TCDD	0.22	
1,2,3,7,8-PeCDD	0.91	
1,2,3,4,7,8-HxCDD	1.4	
1,2,3,6,7,8-HxCDD	0.67	
1,2,3,7,8,9-HxCDD	0.47	
1,2,3,4,6,7,8-HpCDD	2.1	
OCDD	3.7	
OCDF	2.7	
Total toxicity (TEQ-WHO98)	458	

by leading Russian and foreign laboratories (Scientific Production Association "Typhoon", Scientific-Technical Centre of Radiation and Chemical Safety and Hygiene of the Federal Medical Biological Agency of Russia (STC RCSH), A. N. Severtsov Institute of Ecology and Evolution, Kuban State Agrarian University and Federal office of Consumer Protection and Food Safety (BVL)—Germany). The results obtained by ILC participants are shown in Fig. 2. The generalised results of the ILC were presented at the 3rd All-Russian Conference "Analytical Chromatography and Capillary Electrophoresis" [35].

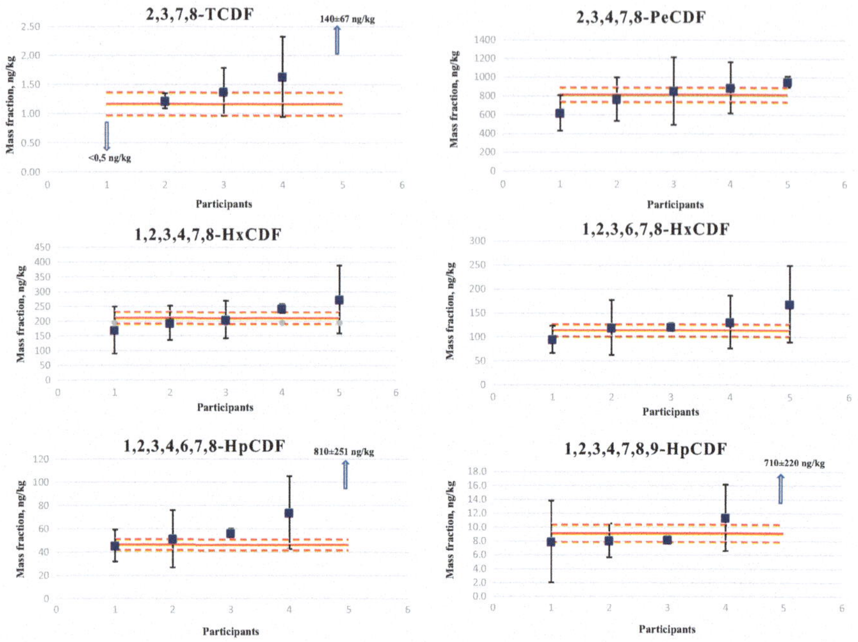

Fig. 2 Measurements of the mass fraction of dioxin congeners during interlaboratory comparison studies. The results are shown in the ascending order of mass fraction values. (The solid red line depicts the certified value; dotted red line is the range of values ± expanded uncertainty. The arrow and the numerical value denote the result that is significantly different from the certified value of the congener) [16]

As can be seen from the data presented in Fig. 2, the major part of the results are within the measurement error characteristic of most chromatographic techniques—up to 40%. Thus, the measurement results obtained by ILC participants can be considered satisfactory.

In general, the ILC results show the mean and median of the measurement results to be in good agreement with the certified values of the CRM.

Additional studies were carried out to estimate the mass fraction of PCDD, the content of which did not exceed the value of 4 ng/g in the CRM. The measurements were performed by GC-MS/MS characterised by a higher (compared to GC—LRMS) selectivity level and better limiting characteristics (detection limit and analyte detection limit). The obtained measurement results will be provided as additional information about the CRM.

Examples of the mass fragmentation patterns of CRM samples obtained via GC—LRMS and GC—MS/MS are shown in Fig. 3.

HxCDD

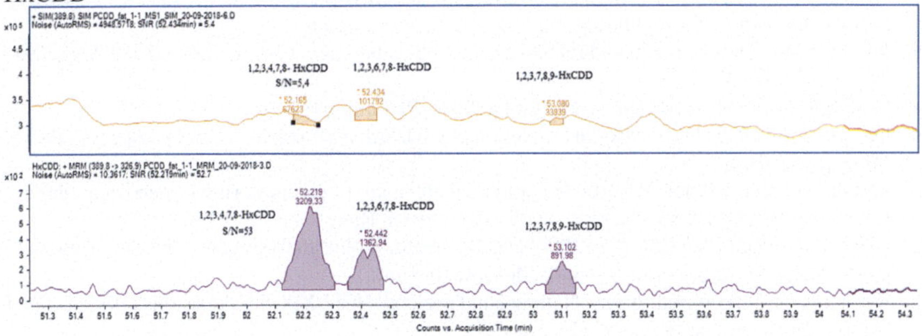

Fig. 3 Mass fragmentation patterns of a CRM sample, reconstructed by ions characteristic of HxCDD, in SIM m/z 390 mode (top) and in MRM m/z 390/327 mode (bottom): 52.219 min—1,2,3,4,7,8-HxCDD; 52.442 min—1,2,3,6,7,8-HxCDD; 53,102 min—1,2,3,7,8,9-HxCDD

4 Conclusion

A full range of studies aimed at determining the dioxin (PCDF) mass fraction in animal fat has been carried out using an RM prepared for the purposes of the research. A methodology and algorithm for the certification of CRM matrices have been developed.

The developed CRM can be used in the certification, validation and verification of measurement methods, as well as in the quality control of measurements performed to estimate the dioxin content level in fat matrices.

Trials aimed at approving the developed CRM of the mass fraction of dioxins in animal fat and submitting the information to the Federal Information Fund for Ensuring the Uniformity of Measurements (FIFEUM) are scheduled for 2019. Afterwards, the CRM is expected to be commercially available for Russian and foreign users.

Acknowledgements The authors express their sincere gratitude to the staff of the Scientific Production Association "Typhoon", Scientific-Technical Centre of Radiation and Chemical Safety and Hygiene of the Federal Medical Biological Agency of Russia (STC RCSH), A. N. Severtsov Institute of Ecology and Evolution, Kuban State Agrarian University and Federal Office of Consumer Protection and Food Safety (BVL), Germany, for their participation in interlaboratory comparisons.

Author Contribution Statement Budko A. G.: Preparation of the reference material, collection and analysis of experimental data, literature review, preparation of the manuscript draft;

Mikheeva A. Y.: Literature review, revision of the manuscript;

Krylov A. I.: Critical analysis and revision of the manuscript.

References

1. The effects of herbicides in South Vietnam. Part A. Summary and conclusion (1974) National Academy of Sciences, Washington

2. Ambient water quality criteria 2,3,7,8-tetrachlorodibenzo-p-dioxin: report EPA-400/5-84/007 (1984) Office of water regulations and standards, Washington
3. Vysochin VI (1989) Dioxin and related compounds: analytical review. GPNTB SO AN SSSR, Novosibirsk (in Russ.)
4. Gough M (1986) Dioxin, agent orange: the facts. Plenum press, NY
5. Fedorov LA (1993) Dioxins as an environmental hazard: a retrospective and prospects. Nauka, Moscow (in Russ.)
6. Maistrenko VN, Klyuev NA (2004) Ecological and analytical monitoring of persistent organic pollutants. BINOM. Laboratoriya znanii, Moscow (in Russ.)
7. Drugov YS, Rodin AA (4th eds) (2006) Gas chromatographic analysis of polluted air: a practical guide. BINOM. Laboratoriya znanii, Moscow (in Russ.)
8. Registries of dioxins and furans. National and regional PCDD/PCDF emissions (1999) United Nations Environment Program, Geneva
9. Stockholm Convention on Persistent Organic Pollutants (2001) Stockholm
10. Pozdnyakov SP, Rumak VS, Sofronov GA, Umanova NV (2006) Dioxins and human health. Nauka, Saint Petersburg
11. TR CU 021/2011 Technical Regulations of the Customs Union "On food safety" (2018). Available via Eurasian Economic Union. https://docs.eaeunion.org/pd/ru-ru/0102880/pd_16052018. Accessed 20 Sept 2018
12. Commission Regulation (EC) No 1881/2006 Setting maximum levels for certain contaminants in foodstuffs. Official J Eur Union 49 (L 364)
13. GOST 31792-2012 (2014) Fish, marine invertebrates and their products. Determination of dioxins and dioxin-like polychlorinated biphenyls by the chromatographic mass-spectrometric method. Standartinform, Moscow (in Russ.)
14. PND F 13.2: 3.64—08 (2014) Methods of measuring the mass concentration of polychlorinated dibenzo-n-dioxins and dibenzofurans in the atmospheric air of the working area using chromatography and mass spectrometry. Moscow (in Russ.)
15. Method 1613 Rev. B (1994) Tetra-through octa-chlorinated dioxins and furans by isotope dilution HRGC/HRMS, U.S. Environmental protection agency office of water engineering and analysis division, Washington, p 89
16. ISO 18073:2004 (2013) Water quality—determination of tetra- to octa-chlorinated dioxins and furans—method using isotope dilution HRGC/HRMS. ISO
17. EN 1948-2 (2006) Stationary source emissions—determination of the mass concentration of PCDDs/PCDFs and dioxin-like PCBs—part 2: extraction and clean-up of PCDDs/PCDFs
18. EN 1948-3 (2006) Stationary source emissions—determination of the mass concentration of PCDDs/PCDFs and dioxin-like PCBs—part 3: identification and quantification of PCDDs/PCDFs
19. JIS K 0311-1999 (1999) Method for determination of tetra-through octa-chlorodibenzo-p-dioxins, tetra-through octa-chlorodibenzofurans and coplanar polychlorobiphenyls in stationary source emissions, Japanese Standards Association
20. Mutual recognition of national measurement standards and calibration and measurement certificates issued by national metrological institutes. CIPM MRA Agreement of 14.10.1999 (1999). Available via ROSSTANDART Federal Information Foundation for ensuring the uniformity of measurements. https://fgis.gost.ru/fundmetrology/registry/7. Accessed 20 Sept 2018 (in Russ.)
21. RMG 55-2003 GSI (2004) Standard samples of the composition of pure organic substances. Methods of testing. Main principles. Izdatel'stvo standartov, Moscow (in Russ.)
22. Duewer DL, Parris RM, White E, May WE, Elbaum H (2004) An approach to the metrologically sound traceable assessment of the chemical purity of organic reference materials. Special Publication NIST SP, 1012
23. Certified Reference Materials catalogue (2018) Joint Research Centre—Institute for Reference Materials and Measurements (JRC-IRMM). https://ec.europa.eu/jrc/en/reference-materials/catalogue. Accessed 20 Sept 2018
24. Approved types of reference materials RF (2018) ROSSTANDART Federal Information Foundation for ensuring the uniformity of measurements, Moscow. https://fgis.gost.ru/fundmetrology/registry/19. Accessed 20 Sept 2018

25. Wellington laboratories current catalogue inc 2016–2018 (2018) Wellington laboratories, Canada. https://well-labs.com/products/catalogue. Accessed 20 Sept 2018
26. Dioxin and furan method standards, standard mixtures, and reference materials (2018) Cambridge Isotope Laboratories. http://shop.isotope.com/category.aspx?id=10032753. Accessed 20 Sept 2018
27. Standard reference materials (2018) National Institute of Standards and Technology. https://www.nist.gov/srm
28. MUK-99 (1999) Guidelines for the identification and isomer-specific determination of polychlorinated dibenzo-p-dioxins and dibenzofurans in meat, poultry, fish, products and by-products from them, as well as in other fat-containing products and feed by chromatography mass spectrometry. Moscow (in Russ.)
29. De Bièvre P (1993) Isotope dilution mass spectrometry as a primary method of analysis. Anal. Proc. https://doi.org/10.1039/ap9933000328
30. Milton MJT, Quinn TJ (2001) Metrologia 38:289–296
31. ILAC-G12:2000 Guidelines for the Requirements for the competence of reference material producers
32. Kaarls R (2018) The consultative committee for metrology in chemistry and biology—CCQM. J Chem Metrol 12(1):1–16. https://doi.org/10.25135/jcm.11.17.12.060
33. GOST ISO Guide 35—2015 (2016) Reference materials—general and statistical principles of certification. Standartinform, Moscow (in Russ.)
34. RMG 93–2015 (2016) GSI Evaluation of the metrological characteristics of certified reference materials. Standartinform, Moscow (in Russ.)
35. Budko AG, Mikheeva AYu, Krylov AI (2017) On the accuracy and reliability of the results of measurement of polychlorinated dibenzo-(p)-dioxins and dibenzofurans. The results of interlaboratory comparisons. In: Materials of the III All-Russian Conference "analytical chromatography and capillary electrophoresis". Kuban State University, Krasnodar, 21–27 May 2017

Development of Certified Reference Materials for Physicochemical Analysis. Certified Reference Solutions for the Composition of Aqueous Aluminium, Indium, Magnesium, Nickel and Titanium

Ianina I. Ermakova, Alexander V. Ivanov, Irina N. Zyablikova and Anna N. Shobina

Abstract This paper presents information on the development and certification of reference materials (CRM) for the composition of aqueous aluminium, indium, magnesium, nickel and titanium solutions. CRMs are represented by aqueous solutions containing aluminium, indium, magnesium, nickel and titanium ions. CRMs have been packed into 5, 10 and 25 cm^3 glass ampoules, as well as into 50 and 250 cm^3 polypropylene vessels. The certified values of metal mass concentrations were established according to the preparation method. The certified value of the mass concentration of metal ions in the CRMs under study is found to vary within the range of 0.95–1.06 g/dm^3. Both the relative expanded uncertainty (under $k = 2$) of the certified mass concentration and the relative standard uncertainty due to instability do not exceed the values of 0.8% and 0.069%, respectively. The developed CRMs are shown to have the shelf life of 3 years provided that standard storage conditions are ensured. The developed CRMs are included into the RF State Register of Certified Reference Materials under the following numbers: CRM 11122-2018, CRM 11123-2018, CRM 11124-2018, CRM 11125-2018 and CRM 11126-2018. The CRMs produced by the All-Russian Scientific Research Institute for Optical and Physical Measurements (VNIIOFI) are characterized by the values of expanded uncertainty comparable to those of world leaders in the production of similar materials.

Keywords Reference material · Spectral methods · GET 196-2015 · State primary standard · Aluminium · Nickel · Magnesium · Titanium · Indium · Chemical analysis

---

I. I. Ermakova (✉) · A. V. Ivanov · I. N. Zyablikova · A. N. Shobina
The All-Russian Scientific Research Institute for Optical and Physical Measurements (VNIIOFI), Moscow, Russia
e-mail: Ermakova@vniiofi.ru

S. Medvedevskikh et al. (eds.), *Reference Materials in Measurement and Technology*,
https://doi.org/10.1007/978-3-030-32534-3_3

Abbreviations

CRM	Certified reference material
RM	Reference material
ICP-OE	Inductively coupled plasma atomic emission spectrometry
ICP-MS	Inductively coupled plasma mass spectrometry
R&D	Research and development

1 Introduction

Reference materials for the chemical composition of metal ion solutions are intended for the determination of the corresponding ions in water, soil, atmospheric air, biological environments, environmental objects, as well as in food, technical and chemical products using photometric, spectrophotometric, atomic absorption, ICP-OE, ICP-MS and various other methods. All these measurement methods are indirect, thus requiring preliminary calibration. Analytical samples used in such devices are mainly aqueous solutions; therefore, these devices are calibrated using calibration standards frequently prepared by mixing and diluting certified reference materials (CRMs) for the composition of solutions, thus forming the necessary system of certified mixtures. Such standard solutions should meet specific requirements determined by the method used and formulated objectives (low detection limits, high determination accuracy).

Standard reference materials are used not only for the calibration of measurement instruments, but also for the metrological certification of measurement procedures and control of their accuracy.

The All-Russian Scientific Research Institute for Optical and Physical Measurements (VNIIOFI) provides metrological support to various scientific and industrial organisations, including those working in aviation, metallurgy and other fields. For this purpose, the Institute has developed and certified a state primary standard of the mass (molar) fraction and mass (molar) concentration of components in liquid and solid substances and materials using spectral methods GET 196-2015.[1] This standard is used for the development and certification of reference measurement methods and reference materials, as well as for the establishment of metrological traceability. Stricter requirements for the accuracy of measurements call for more accurate reference materials with the certified value uncertainty lower than 1%.

[1]GET 196-2015—RF State primary standard of units of mass (molar) fraction and mass (molar) concentration of components in liquid and solid substances and materials on the basis of spectral methods. Available via VNIIOFI. http://www.ckp.vniiofi.ru/index.php/16-etalony/41-get-196-2011-gosudarstvennyj-pervichnyj-etalon-edinits-massovoj-molyarnoj-doli-i-massovoj-molyarnoj-kontsentratsii-komponentov-v-zhidkikh-i-tverdykh-veshchestvakh-i-materialakh-na-osnove-spektralnykh-metodov.

1.1 Literature Review

For the fabrication of the CRMs for the composition of metal solutions in VNIIOFI, five metals were selected, including aluminium, nickel, magnesium, titanium and indium. These metals were chosen as the most common elements present in the matrices of major aircraft-grade alloys. Indium is also widely used in spectral analysis as the internal standard, since its pronounced spectral lines do not overlap with those of other elements in samples under study.

Prior to experiments, information on existing reference materials of the selected type was analysed. The literature review has shown that the Federal Information Fund for Ensuring the Uniformity of Measurements currently lacks certified reference materials for aqueous indium solutions.

It can be seen from Tables 1, 2, 3 and 4 that some reference materials were developed using the respective metal salts as the carriers of the component under certification. As a result, the combined uncertainty of the developed CRMs will include the uncertainty of weighing, as well as that of the determination of the solubility loss of the deposited compounds and the determination of the concentration of co-deposited impurities. Eventually, the combined relative uncertainty value can vary within the range of ±1.0% [1–3].

2 Materials and Methods

In VNIIOFI, reference materials (RMs) for the chemical composition of metal solutions were prepared by dissolving high-purity metals in nitric, hydrochloric and hydrofluoric acids. Starting materials used for the development of RMs were all produced in Russia. The mass fraction of the main component in these materials was calculated as the difference between 100% and the total impurity content. The identification of the purity of the certified element carrier for aluminium, nickel, magnesium and titanium was carried out in 2015 under the auspices of the research project "Research in the field of measuring the physico-chemical composition and properties of substances in the development of primary state standards as high-purity substances for the reproduction of physical quantities that characterize the chemical composition of solids", code *Chistota*. Chemically pure metals were purchased from the *Komponent-Reaktiv* Ltd. company (Moscow, Russia) and investigated using the GET 196-2015. The resulting values of mass fraction and uncertainty are presented in Table 5.

For the fabrication of an indium CRM, In000 prepared according to GOST 10297-94 [4] purchased from the *Giredmet* company (RF), was used. The mass fraction of indium in In000 equals 99.9995%. Aluminium, nickel, magnesium, titanium and indium ion solutions were prepared using the following equipment and materials:

- Laboratory balances *Sartorius* AG ME36S (Germany), special accuracy class (1) according to GOST R 53228-2008 [5];

Table 1 CRMs for the composition of aqueous aluminium solutions

CRM number	Designation	Manufacturer	Starting material	Certified value of aluminium ion mass concentration	Permissible values of relative uncertainty (%)
8059-94/8061-94	RM for the composition of aqueous aluminium ion solutions (set no. 12K)	Centre of Reference Materials and High-Purity Substances Ltd., RF	Aqueous solutions of potassium alums acidified with sulphuric acid	0.95–1.05	±1.0
7927-2001	RM for the composition of aluminium ion solutions (NK-EK)	Ekroschim Ltd., RF	Solution of aluminium in 1 M of nitric acid	0.95–1.05	±1.0
7854-2000	RM for the composition of aluminium ion solutions (42K)	Centre of Reference Materials and High-Purity Substances Ltd., RF	Aqueous solution of aluminium nitrate acidified with nitric acid	0.95-1.05	±1.0
7758-2000	RM for the composition of aluminium ion solutions	Ecological Analytical Association "Ekoanalytica", RF	Solution of aluminium-potassium alums in 1 M of sulphuric acid	0.95–1.05	±1.0
7453-98	RM for the composition of aluminium ion solutions (set 1K)	D.I. Mendeleyev Institute for Metrology (VNIIM), RF	Aluminium chloride hexahydrate (produced by the *Panreac* company) solution in deionized water acidified to pH 2–3	0.475–0.525 0.95–1.05	±3.0
7269-96	RM for the composition of aluminium ion solutions	Ural Scientific Research Institute for Metrology (UNIIM), Ural Plant of Chemical Reagents, RF	Alum solution in nitric acid	0.95–1.05	±1.0

Table 2 CRMs for the composition of aqueous nickel solutions

CRM number	Designation	Manufacturer	Starting material	Certified value of nickel ion mass concentration	Permissible values of relative uncertainty (%)
8001-93/8003-93	RM for the composition of aqueous nickel ion solutions (set no. 6K)	Centre of Reference Materials and High-Purity Substances Ltd., RF	Aqueous solutions of nickel (II) nitrate, acidified with nitric acid	0.95–1.05	±1.0
7873-2000	RM for the composition of nickel ion solutions (NK-EK)	Ekroschim Ltd., RF	Solution of nickel in 1 N nitric acid	0.95–1.05	±1.0
7785-2000	RM for the composition of nickel ion solutions	Ecological Analytical Association "Ekoanalytica", RF	Solution of nickel nitrate in 1 M of nitric acid	0.95–1.05	±1.0
7442-98	RM for the composition of nickel ion solutions (set 11K)	D.I. Mendeleyev Institute for Metrology (VNIIM), RF	Nickel nitrate solution in deionized water acidified to pH 2–3	0.475–0.525 0.95–1.05	±1.0
7265-96	RM for the composition of nickel ion solutions	Ural Scientific Research Institute for Metrology (UNIIM), Ural Plant of Chemical Reagents, RF	Solution of 7-hydrate nickel sulphate without cobalt in nitric acid	0.95–1.05	±1.0
7111-94	RM for the composition of nickel (II) ion solutions	Ekmets Ltd., RF	Solution of nickel (II) ions in 0.1 M nitric acid solution	0.95–1.05	±1.0

Table 3 CRMs for the composition of aqueous magnesium solutions

CRM number	Designation	Manufacturer	Starting material	Certified value of magnesium ion mass concentration	Permissible values of relative uncertainty (%)
7767-2000	RM for the composition of magnesium ion solutions	Ecological Analytical Association "Ekoanalytica", RF	Solution of magnesium nitrate in 0.1 M of nitric acid	0.95–1.05	±1.0
7681-99	RM for the composition of aqueous magnesium ion solutions	Ekroschim Ltd., RF	Aqueous solution of hexahydrate magnesium chloride	0.95–1.05	±1.0
7445-98	RM for the composition of magnesium ion solutions (set no. 7K)	D.I. Mendeleyev Institute for Metrology (VNIIM), RF	Solution of magnesium sulphate in deionized water, acidified to pH 2–3	0.95–1.05 4.75–5.25	±1.0
7190-95/7192-95	RM for the composition of aqueous magnesium ion solutions (set no. 20K)	Centre of Reference Materials and High-Purity Substances Ltd., RF	Aqueous solutions of magnesium chloride	0.95–1.05 0.475–0.525 0.095–0.105	±1.0

- Volumetric polypropylene flasks with a capacity of 50 cm^3, 1st accuracy class, or individually calibrated with an uncertainty of no higher than ±0.5% of the nominal volume;
- Automatic dispensers *Biohit mLine* (Finland), State Register number 43133-09, individually calibrated;
- Water distilling apparatus according to the specification 64-1-1640-72;
- Device for obtaining ultrapure water *Vodoley* (Russia) according the specification 2.115.000.000;
- Acid sub-boiling purification system *BERGHOF BSB-939-IR* (Germany);
- Purified hydrochloric acid according to GOST 3118-77 [6] (purified by sub-boiling acid distillation);
- Purified nitric acid according to GOST 4461-77 [7] (purified by sub-boiling acid distillation);

Table 4 CRMs for the composition of aqueous titanium solutions

CRM number	Designation	Manufacturer	Starting material	Certified value of titanium ion mass concentration	Permissible values of relative uncertainty (%)
8464-2003	RM for the composition of titanium (IV) ion solutions	Ecological Analytical Association "Ekoanalytica", RF	Titanium sulphate solution (IV) in 0.5 M sulphuric acid	0.95–1.05	±1.0
7205-95/7207-95	RM for the composition of aqueous titanium (IV) ion solutions (set no. 13K)	Centre of Reference Materials and High-Purity Substances Ltd., RF	Titanium tetrachloride aqueous solution, acidified with sulphuric acid	0.95–1.05 0.475–0.525 0.095–0.105	±1.0

Table 5 Metrological characteristics of pure metals

Certified characteristic	Certified value (%)	Expanded uncertainty of the certified value $U_{(k=2)}$ (%)
Nickel mass fraction	99.996	0.0013
Titanium mass fraction	99.877	0.0118
Aluminium mass fraction	99.9888	0.0030
Magnesium mass fraction	99.9287	0.0082

– Purified hydrofluoric acid according to GOST 2567-89 [8] (purified by sub-boiling acid distillation).

The purity of acids, water and laboratory equipment was controlled using an atomic absorption spectrometer *AA 280Z* (Varian, Australia), serial No. EL07093128, State Registry number 16496-09, included into the RF State primary standard of units of mass (molar) fraction and mass (molar) concentration of components in liquid and solid substances and materials on the basis of spectral methods GET 196-2015.

The solutions of aluminium, nickel, titanium, magnesium and indium were prepared from the respective metal swarf taken in the amount of 0.5 g. Swarf portions were weighted using a laboratory electronic balance *Sartorius* ME 36S (Germany) and subsequently placed into the autoclaves of a *DigiPREP* sample preparation system (SCP Science, Canada). The obtained samples were dissolved in 1 M of hydrochloric acid (for aluminium), 1 M of nitric acid (for magnesium, indium and nickel), 1 M of nitric acid and 0.2 M of hydrofluoric acid (for titanium) at a maximal temperature of 100 °C. The acids were previously purified by distillation using a

Berghof BSB-939-IR acid purification system. The obtained solutions with the metal ion mass concentration of 10 g/dm^3 underwent 10-fold dilution by ultrapure water according to GOST 6709-72 [9].

The certified values of metal ion mass concentrations were calculated according to the preparation procedure.

3 Investigation of RM Instability

The instability of the developed RMs was investigated in accordance with the RMG 93-2015 interstate standard [10] by the isochronous method using an atomic emission installation included into the GET 196-2015. In the isochronous stability research, the accelerated aging method was applied. The samples were divided into two groups. One group was stored at (20 ± 5) °C, whereas the other group was stored at elevated temperatures. At fixed storage temperatures, the duration of the stability research is estimated by the formula:

$$\tau = \frac{T}{2^{\frac{t_1 - t_0}{10}}}, \quad (1)$$

where T is the expected shelf life of an RM sample equalling 36 months; t_0, t_1—the temperatures of RM storage under normal conditions and under accelerated aging, respectively.

Under the RM storage temperature of (20 ± 5) °C and the accelerated aging temperature of 80 °C,

$$\tau = \frac{12}{2^{\frac{80-20}{10}}} = 0.56. \quad (2)$$

The storage of an RM sample during 403 h at a temperature of 80 °C corresponds to 36 months at 20 °C. During 403 h every three days, five pairs of measurements were carried out under repeatability conditions, with the measurement deviation d_i at the i-th time point being calculated.

The standard uncertainty due to instability u_{stab} at the t time point and the number of degrees of freedom v_{stab} were assessed according to Formula 3. Below is the calculation performed for aluminium.

$$u_{stab} = s_a t = 0.069\%, \quad (3)$$

where t is the RM shelf life of 3 years; s_a is the standard deviation of the linearity coefficient in the dependency of the measurement deviation d_i on time.

No statistically significant alteration during the period of the stability research was found.

4 Calculation of the Expanded Uncertainty of Certified Values

The uncertainty associated with the measurement method used for determining RM certified values was calculated in accordance with the GOST ISO Guide 35-2015 [11] and RMG 93-2015. The certified value was taken to be the calculated value of metal mass concentration obtained by the preparation method.

The standard uncertainty associated with the characterization method was estimated by Formula 4. The calculation below is given for aluminium.

$$u_{char} = \sqrt{\frac{s^2}{n-1} + \frac{U^2(\Theta)}{k^2}} = 0.267\%, \quad (4)$$

where $U(\Theta)$ is the expanded standard uncertainty arising from the preparation method; s is the relative standard deviation of the prepared parallel RMs.

The combined standard uncertainty of the RM certified values according to GOST R 54500.3-2011 was determined by the following equation:

$$u_c^2(A) = u_{char}^2 + u_{stab}^2 = 0.276\%, \quad (5)$$

where u_{char}^2 is the standard uncertainty associated with the determination method of RM certified values; u_{stab}^2 is the standard uncertainty associated with RM instability.

The expanded uncertainty U was calculated by multiplying the combined standard uncertainty u_c by the coverage factor $\boldsymbol{k}$ using the formula

$$U = ku_c = 0.55\%. \quad (6)$$

5 Results and Discussion

The results of the experiments carried out taking into account the stability of the obtained solutions and the uncertainty associated with the determination method are summarised in Table 6. It presents the mass concentrations of metal ions and the expanded uncertainty of their certified values.

The description of the developed reference materials includes the expanded uncertainty of 0.8%, which was calculated taking into account the maximum possible values of all uncertainty components, including the method used for determining certified characteristics, purity of materials, etc.

The certified reference materials for the chemical composition of the aluminium, titanium, magnesium, nickel and indium ion solutions fabricated by *VNIIOFI* are packed into ampoules with a capacity of 5, 10 and 25 cm^3, as well as into polypropylene vessels with a capacity of 50 and 250 cm^3.

Table 6 Certified reference materials (CRMs) for the composition of the aqueous metal ion solutions developed by *VNIIOFI*

CRM number	Designation	Starting material	Certified value of the mass concentration of metal ions		Expanded uncertainty of the certified value under the coverage factor k = 2 (%)	
			According to the description	First batch	According to the description	First batch
11122-2018	RM for the composition of aluminium solutions	Solution of metallic aluminium in 1 M of hydrochloric acid	0.95–1.06	1.04	0.8	0.55
11123-2018	RM for the composition of indium solutions	Solution of metallic indium in 1 M of nitric acid	0.95–1.06	1.03	0.8	0.31
11124-2018	RM for the composition of magnesium solutions	Solution of metallic magnesium in 1 M of nitric acid	0.95–1.06	0.99	0.8	0.44
11125-2018	RM for the composition of nickel solutions	Solution of metallic nickel in 1 M of nitric acid	0.95–1.06	0.99	0.8	0.37
11126-2018	RM of the composition of titanium solutions	Solution of metallic titanium in 1 M of nitric acid and in 0.2 M of hydrofluoric acid	0.95–1.06	1.05	0.8	0.52

6 Conclusion

The developed reference materials have been included into the RF State Register of Certified Reference Materials under the following numbers: CRM 11122-2018, CRM 11123-2018, CRM 11124-2018, CRM 11125-2018, CRM 11126-2018.[2] The distinctive feature of the CRMs manufactured by *VNIIOFI* consists in the statement

[2]CRM 11122-2018 Reference material for the composition of aluminium solutions. Rosstandard. https://fgis.gost.ru/fundmetrology/registry/19/items/492841.

of expanded uncertainty (rather than error) values in their metrological characteristics. This facilitates integration of Russian metrology into international metrological organizations, such as *COOMET, BIPM, JCTLM*.

The certified value of the mass concentration of metal ions in the CRMs is found to vary within the 0.95–1.06 g/dm^3 range. The relative expanded uncertainty ($k = 2$) of the certified value of mass concentration does not exceed 0.8%, with the relative standard uncertainty caused by instability being lower than 0.069%. The shelf life of the RMs is established to be 3 years provided that storage conditions are ensured.

The CRMs manufactured by *VNIIOFI* are characterized by the values of expanded uncertainty comparable to those of world leaders in the production of similar materials.

Meeting the required accuracy characteristics, the developed CRMs of metal ions have been included into the RF State primary standard of units of mass (molar) fraction and mass (molar) concentration of components in liquid and solid substances and materials on the basis of spectral methods GET 196-2015.

Author Contribution Statement Ermakova Y. I.: Research concept, collection of experimental data, analysis of experimental information, literature review, manuscript preparation, translation into English, critical analysis and revision of the manuscript.

Zyablikova I. N.: Collection of experimental data, analysis of experimental data, preparation of documents for certification of the obtained RMs.

Shobina A. N.: Development of the article concept, analysis of existing literature, preparation of the manuscript draft, critical analysis and revision of the manuscript.

Ivanov A. V.: Concept of research, critical analysis and revision of the article.

References

1. Laitinen HA, Harris VE (1979) Chemical analysis. Chemistries, Moscow (in Russ.)
2. Babko AK, Pyatnitsky NV (1968) Quantitative analysis. Mir, Moscow (in Russ.)
3. Lisienko DG, Dombrovskaya MA, Lisienko MD (2010) Certified reference materials for composition of metal ion solutions for calibration of instruments with inductively coupled plasma. Ref Mater 3:45–56 (in Russ.)
4. GOST 10297-94 (1994) Indium. Specifications. Minsk, Interstate Council on Standardisation, Metrology and Certification, p 27 (in Russ.)
5. GOST 53228-2008 (2010) Non-automatic weighing instruments. Standartinform, Moscow, 134 p (in Russ.)
6. GOST 3118-77 (2018) Reagents. Hydrochloric acid. Specifications. IPK publishing standards, Moscow, 12 p (in Russ.)

CRM 11123-2018 Reference material for the composition of indium solutions. Rosstandard. https://fgis.gost.ru/fundmetrology/registry/19/items/492842.

CRM 11124-2018 Reference material for the composition of magnesium solutions. Rosstandard. https://fgis.gost.ru/fundmetrology/registry/19/items/492843.

CRM 11125-2018 Reference material for the composition of nickel solutions. Rosstandard. https://fgis.gost.ru/fundmetrology/registry/19/items/492844.

CRM 11126-2018 Reference material for the composition of titanium solutions. Rosstandard. https://fgis.gost.ru/fundmetrology/registry/19/items/492845.

7. GOST 4461-77 (2018) Reagents. Nitric acid. Technical conditions. Standartinform, Moscow. 6 p (in Russ.)
8. GOST 2567-89 (2018) Hydrofluoric acid technical. Technical conditions. IPK publishing standards, Moscow, 15 p (in Russ.)
9. GOST 679-72 (2016) Water distilled. Specifications. Standartinform, Moscow, 11 p (in Russ.)
10. RMG 93-2009 GSI (2011) State system for ensuring the uniformity of measurements. Estimation of metrological characteristics of reference materials. Standartinform, Moscow, 30 p (in Russ.)
11. GOST ISO Guide 35-2015 (2016) Reference materials—basic and statistical principles of the certification. Standartinform, Moscow, 61 p (in Russ.)

Multiparametric Reference Materials for Metrological Support in the Area of Coating Parameter Control

Vyacheslav V. Kazantsev and Alexander S. Vasilyev

Abstract The present work analyses the state and development prospects of metrological support and standardisation in the field of measuring surface density, thickness and chemical composition of coatings. The main characteristics of the State Primary Measurement Standard GET 168-2015 and developed multiparametric reference materials are described.

Keywords Multiparametric reference material · Coating · Thickness · Surface density · Chemical composition · Verification scheme · Error · Uncertainty

Abbreviations

RM	Reference material
XRF	X-ray fluorescence
GET, GET 168-2015	State Primary Measurement Standard for units of surface density and mass fraction of elements in coatings 168-2015
TS	Transfer standard

1 Introduction

Over the last decade, there has been a tendency to develop instruments for the universal measurement of coating parameters based on the X-ray fluorescence (XRF) analysis method, allowing not only the parameters of single-layer and single-component coatings to be measured, but also of multilayer and multicomponent coatings where both thickness and surface density are to be determined [1].

V. V. Kazantsev (✉) · A. S. Vasilyev
Ural Scientific Research Institute for Metrology (UNIIM), Ekaterinburg, Russia
e-mail: kazantsev@uniim.ru

S. Medvedevskikh et al. (eds.), *Reference Materials in Measurement and Technology*,
https://doi.org/10.1007/978-3-030-32534-3_4

This fact is confirmed by the type-approval testing and submission of more than 10 types of instruments, XRF analysers and coating thickness gauges to the Federal Information Fund for Ensuring the Uniformity of Measurements. These instruments have comprehensive measuring capabilities, including the ability to measure the chemical composition of multicomponent and multilayer (up to 24 layers) coatings, surface density from 0.1 to 500 g/m^2, thickness of coatings from 0.01 to 50 μm and chemical composition from 1 to 100%.

At the same time, metrological support in this field of measurement—above all, the creation of standard measuring instruments, including the State Primary Measurement Standard GET 168-2015 [2], RMs for surface density, thickness and chemical composition of two-component coatings [3, 4]—was being developed. This indicates the need for a comprehensive solution to the problem of ensuring the uniformity of coating parameter measurements, considering the relevance of this field of measurement in the development and implementation of modern technologies for creating special—including composite and nanostructured—coatings, as well as the need to measure and control the parameters of such coatings.

2 Literature Review

It should be noted that XRF, which has certain advantages over other methods—and primarily because it is non-destructive—has received a lot of attention in the scientific literature. A representative review of works related to the use of XRF, mainly for carrying out quantitative chemical analysis of various solid-state films and coatings, is given [5]. Of particular interest here is a cycle of studies on manufacturing processes, including experimental and theoretical studies of systems in the form of two-component Co–Ni/Cr [6] and two-layer films [7]. The results of studies of bilayer films on polycore (nonporous transparent ceramics) are provided in [7]. These films are obtained using the ion-plasma sputtering method in a vacuum. The top three-component (Fe–Ni–Mo) layer of these films has a thickness ranging from 300 to 430 nm, while the bottom layer is formed of a single component (Cr) having a thickness from 27 to 50 nm. Regulatory support in this type of measurement is represented by national [8, 9] and foreign standards [10–12].

The State verification scheme for measuring the surface density of coatings, regulated by the national standard [8] was revised to meet the new metrological characteristics of GET 168-2015. This standard has been replaced by the "State verification schedule for instruments measuring the surface density of coating" since 1st January 2019 according to the order of Rosstandart No. 2089 of 28th September 2018. This verification scheme supports two branches for unit transfer: one ("more accurate") for single-component and multicomponent single-layer coatings, the other ("less accurate") for multilayer coatings.

The national standard [9] establishes the classification, application procedure and execution of documents for the verification of coating thickness gauges in the form of thickness measurements and RMs for surface density of coatings.

International standards establish requirements for the measurement capabilities of the XRF method, including measurement range [10], terminology, basic measurement principles, calibration and evaluation of measurement results using the XRF method [11, 12].

3 Materials and Methods

In 2016–2018, the Ural Scientific Research Institute for Metrology (UNIIM) conducted research of two types of RMs for surface density, thickness and mass fraction of elements in coating using GET 168-2015:

- CRM with an iron-nickel coating applied on silicon base (GSO 10880-2017);
- CRM with an iron-nickel-cobalt coating applied on silicon base (pending approval).

The base of these RMs consists of monocrystalline silicon in the form of a plate having a diameter of 60 ± 0.5 mm and thickness of 0.3 ± 0.1 mm. The nickel-iron (or iron-nickel-cobalt) coating was applied on a specially prepared plate by the high-frequency ionic-plasma magnetron sputtering method using an Orion 8 sputtering system (AJA International, USA) at Institute of Natural Sciences, Ural Federal University (Ekaterinburg) at the Department of Magnetism and Magnetic Nanomaterials.

Measurement of the surface density, thickness and mass fraction of iron and nickel in the coating of GSO 10880-2017 was carried out according to the developed measurement procedure MV-07-GET-168-2015 [13]. The uncertainty was estimated according to the measurement procedure MRN-10-GET-168-2015 [14] and the calculation of the error was carried out according to the standard RMG 53-2002 [15].

The same procedure was used to produce Transfer standards (TS). TS consist of a silicon base coated with an iron-nickel (or iron-nickel-cobalt) alloy. The entire area of the TS was measured using Unit 5 of GET 168-2015 to obtain integral intensity values of Kα line for iron and nickel (iron, nickel and cobalt).

The integral values of the surface density of TS coatings were determined by the gravimetric method using a CCE66 "Sartorius" Mass Comparator (Germany) and an IMC-50 × 150 Measuring Microscope (Novosibirsk Instrument-Making Plant, Russia). Measurements of the chemical composition of the TS coatings were performed using the State Secondary Standard of units of mass fraction and mass (molar) concentration of metals in liquid and solid substances and materials (GVET 196-1-2012), which is based on the atomic emission spectrometry method according to the measurement procedure MVV-GVET 196-1-2012 [16].

The equation for measuring the mass fraction of elements in the RMs coating is described by the formula

$$c_j = \frac{\widehat{\beta_j} I_j}{\sum_{i=1}^{n} \widehat{\beta_i} I_i}, \quad (1)$$

where $\widehat{\beta_j}$ is the sensitivity coefficient for the j-th chemical element of the coating, determined by the measurement results of the element mass fraction in the TS coatings using GVET 196-1-2012, %/cps;

I_j is the intensity of the j-th chemical element Kα line when measuring RM, cps.

The equation for measuring the surface coating density of RMs is described by the formula

$$\rho = \hat{\rho} \frac{\sum_{i=1}^{n} \widehat{\beta_i} I_i}{\sum_{i=1}^{n} \hat{C}_i}, \quad (2)$$

where $\hat{\rho}$ is the integral value of TS surface coating density, calculated by the gravimetric method, g/m^2.

The thickness of RMs coating in nanometers was calculated by the formula

$$h = \frac{\rho}{\gamma} \cdot 10^9, \quad (3)$$

where γ is the density value of the material coating according to GOST 9.302-88 [17], g/m^3.

The homogeneity of GSO 10880-2017 coating was evaluated by the mass fraction of elements and the surface thickness according to GOST 8.531-2002 [18], while the results of the homogeneity assessment were considered in the error and uncertainty of the RMs certified values.

4 Research Results

The certified characteristics of GSO 10880-2017 are given in Table 1.

As can be seen from Table 1, the characteristics of the error and the expanded uncertainty of the RMs are reduced by 2–3 times with an increase in surface density and thickness of coatings; this is primarily due to an increase in the accuracy of the TS coating mass measurements by the gravimetric method. For the same reason, the accuracy of certified values of the element mass fractions increases, especially in the case of iron.

The characteristics of the RMs for surface density, thickness and mass fraction of elements in the iron-nickel-cobalt coatings applied on silicon, which were obtained during research, are shown in Table 2.

As can be seen from Table 2, the characteristics of accuracy change with an increase in surface density and coating thickness in a similar way to the two-component coating. A noticeable increase by 2–3 times in relation to iron is observed

Table 1 Certified characteristics of GSO 10880-2017

CRM index in the set	Name of certified characteristics	Certified value	Confidence limits of relative error at $P = 0.95$ (%)	Relative expanded uncertainty, under $k = 2$ (%)
1	Surface density of coating (g/m^2)	0.823	2.4	2.2
2		3.372	0.9	1.0
3		8.575	0.8	0.8
1	Thickness of coating (nm)	93.8	2.4	2.2
2		384.5	0.9	1.0
3		977.8	0.8	0.8
1	Mass fraction of iron (%)	12.47	4.6	4.6
2		11.79	1.7	1.8
3		11.76	1.9	1.9
1	Mass fraction of nickel (%)	88.30	0.9	0.9
2		88.13	0.3	0.3
3		87.95	0.4	0.4

Table 2 Test results of RMs having an iron-nickel-cobalt based coating applied on a silicon base

CRM index in the set	Name of certified characteristics	Certified value	Confidence limits of relative error at $P = 0.95$ (%)	Relative expanded uncertainty, under $k = 2$ (%)
1	Surface density of coating (g/m^2)	0.730	1.4	1.5
2		2.84	0.9	1.0
3		8.25	0.9	0.9
1	Thickness of coating (nm)	83.2	1.5	1.6
2		323	0.9	1.0
3		941	0.8	0.9
1	Mass fraction of iron (%)	11.78	2.4	2.5
2		13.92	1.9	2.0
3		11.64	1.9	1.9
1	Mass fraction of nickel (%)	87.00	0.3	0.3
2		84.18	0.3	0.3
3		87.09	0.4	0.4
1	Mass fraction of cobalt (%)	0.86	6.0	6.0
2		1.02	5.0	5.0
3		1.00	6.0	6.0

for the confidence limits of the relative error and the correspondingly increased uncertainty for the mass fraction of cobalt. This is due to the low concentration of this element in the coating.

5 Conclusions

In general, the obtained values of uncertainty and error for the studied RMs correspond to the values specified in the "State verification scheme for instruments measuring surface density and mass fraction of elements in coatings" both for surface density and mass fractions of iron, nickel and cobalt in coatings applied on silicon.

Taking into account the high corrosion resistance properties of the nickel-iron alloy, its high mechanical strength, as well as the well-known stability of the coating (an analogue of GSO 9936-2011) [4], the monitoring frequency of the metrological characteristics of GSO 10880-2017 is set at once every 2 years.

Author Contribution Statement Alexander S. Vasilyev: conducted experimental studies and prepared materials for the article.

Vyacheslav V. Kazantsev: reviewed the literature, summarised the results of the study and prepared a conclusion.

References

1. Kazantsev VV et al (2017) The state and prospects of development of metrological support and standardization in the field of non-destructive testing of coatings by radiation methods. NDT World 20(1):30–31 (In Russ.)
2. Kazantsev VV et al (2018) State primary standard for units of surface density and mass fraction of elements in coatings GET 168–2015. Measur Tech 9:17–19 (In Russ.)
3. Kazantsev VV et al (2014) Investigation of methods and means of creating multiparameter standard samples of the composition and properties of coatings. Ref Mater 1:42–46 (In Russ.)
4. Sobina EP et al (2012) Creating standard samples of the composition, thickness and surface density of permalloy nano-coatings on silicon. Ind Lab. Diagn Mater 78(8):64–68 (In Russ.)
5. Ignatov YuA et al (2011) X-ray fluorescence analysis of solid films and coatings. Anal Control 15(2):126–140 (In Russ.)
6. Machines NI et al (2000) Determination of the composition and thickness of films of the Co–Ni/Cr system using the X-ray fluorescence method. Appl Spectrosc 55(6):689–691 (In Russ.)
7. Machines NI et al (2009) Determining the composition and thickness of two-layer Fe–Ni–Mo/Cr systems/polycor by X-ray fluorescence method. Bull Nizhny Novgorod Univ NI Lobachevsky 2:89–95 (In Russ.)
8. GOST R 8.612–2011 (2014) State system for ensuring the uniformity of measurements. State verification schedule for means of measuring the surface density of coatings in the range from 0.0001 to 1.000 kg/m^2. Moscow, Standartinform (In Russ.)
9. GOST R 55613-2013 (2015) Non-destructive control. Measures exemplary for calibration of thickness gauges of inorganic coatings. General provisions. Moscow, Standartinform (In Russ.)

10. ASTM B568-98 (2014) Standard X-ray spectrometric method for measuring coating thickness
11. DIN 51418-2-2015 X-ray emission and X-ray fluorescence analysis. Part 2. Terms and definitions and basic principles relating to measurement, calibration and evaluation of results
12. ISO 3497:2000 Metallic coatings—Measurement of coating thickness—X-ray spectrometric methods
13. MV-07-GET-168-2015 (2015) State system for ensuring the uniformity of measurements. The procedure for reproducing the units of surface density and mass fraction of elements in coatings using GET 168 (In Russ.)
14. MRN-10-GET-168-2015 (2015) State system for ensuring the uniformity of measurements. The method of calculating the uncertainty of reproducing the units of surface density and the mass fraction of elements in coatings on the GET 168 (In Russ.)
15. RMG 53-2002 (2012) State system for ensuring the uniformity of measurements. Reference materials. Evaluation of metrological characteristics with the use of measurement standards and reference devices (In Russ.)
16. MVV-GVET 196-1-2012 (2012) State system for ensuring the uniformity of measurements. The procedure for reproducing the unit of the mass fraction of the element in the coatings using GVET 196-1-2012 (In Russ.)
17. GOST 9.302-88 (1990) Unified system of protection against corrosion and ageing. Metal and non-metal inorganic coatings. Control methods Moscow, Publishing house of standards (In Russ.)
18. GOST 8.531-2002 (2002) Reference materials of composition of solid and disperse materials. Ways of homogeneity assessment. Moscow, Publishing house of standards (In Russ.)

Reference Materials Used for Specific Safety Evaluation of Human Immunoglobulin and Human Albumin Products: Features of Development, Certification and Application

Olga G. Kornilova, Maxim A. Krivykh, Rauza A. Volkova and Igor V. Borisevich

Abstract Human immunoglobulin and human albumin products can negatively affect patients' health. The residual content of haemagglutinins, anti-D antibodies, prekallikrein activator, as well as the level of anticomplementary activity, are controlled during both production and testing for compliance with normative documentation. Methods for their quantitative evaluation are based on the effects of erythrocytes hemolysis or agglutination and on the amidolytic product properties of cascade reactions. These methods require the obligatory use of reference materials (RMs). The development of RMs, acting as the carrier of the quantitative characteristics of the human immunoglobulins and human albumin impurities, is fraught with the difficulty of selecting a candidate, substantiating the certification methodology and determining the relevant characteristics. In this research, we used human immunoglobulin solutions with normalized haemagglutinin content, anti-D antibodies or anti-complementary activity in a different range, whose quality was studied using haemagglutination and chromogenic methods and the complement fixation test. Methodological approaches to the RM development and certification were justified using system analysis and forensic audit methods. The necessity of using RMs to assess the stability of analytical work by applying anti-complementary activity determination, haemagglutinin content or anti-D antibody methods is demonstrated. It is established that RMs allow the quantitative content of haemagglutinins, anti-D antibodies and prekallikrein activator to be established. The value of the CRM characteristics should be approximated to the qualitative composition of these impurities in the analysed products, but in amounts not detectable (to produce negative RM components) or exceeding permissible maxima (to produce positive RM components).

Keywords Quality control · Human immunoglobulin products · Human albumin products · Specific safety · Anticomplementary activity · Anti-A and anti-B haemagglutinins · Anti-D antibodies · Prekallikrein activator · Reference

---

O. G. Kornilova (✉) · M. A. Krivykh · R. A. Volkova · I. V. Borisevich
Scientific Centre for Expert Evaluation of Medicinal Products, Moscow, Russia
e-mail: Kornilova@expmed.ru

S. Medvedevskikh et al. (eds.), *Reference Materials in Measurement and Technology*,
https://doi.org/10.1007/978-3-030-32534-3_5

materials · Haemagglutination methods · Complement fixation test · Chromogenic assay

Abbreviations

ACA	Anticomplementary activity
PKA	Prekallikrein activator
HA	Human albumin
BRP	Biological reference preparation
WHO	World Health Organisation
HAE	Haemagglutinins
SPh RF	State Pharmacopoeia of the Russian Federation
Ph.Eur.	European Pharmacopoeia
HIG	Human immunoglobulins
IU	International units
IHA test	Indirect haemagglutination test
DHA test	Direct haemagglutination test
PSF	Product specification file
NIBSC	National Institute for Biological Standards and Control
IRM	Industry reference material
GM	General monograph
CFT	Complement fixation test
RM	Reference material

1 Introduction

High requirements for specific safety associated with the effect of drugs on the complement, kallikrein-kinin and haemostasis systems are imposed on the infusion of HIG and HA products. The presence of HAG (IgG anti-A and anti-B antibodies) in a titre of less than 1:64 reduces the risk of extravascular unwanted reactions caused by macrophage activation in the first 10 days after intravenous administration of the HIG preparation to patients of non-I (0) blood type. In cases when the concentration of anti-D antibodies in these products is not more than 1:8 in Rh-positive patients with impaired immune systems, the likelihood of haemolytic complications of immunoglobulin therapy is negligible. An ACA of not more than 50% (which corresponds to no more than 1 CH_{50} per 1 mg of immunoglobulin protein) allows such undesirable reactions as hot flashes, headache, fever, tachycardia to be eliminated. When the content of the PKA in the infusion products of HIG and HA is less than 35 IU/mL, the hypotensive effect associated with the effect of the activated Hagemann β-fragment factor on the kallikrein-kinin system during infusions is minimal. “Anti-A and anti-B haemagglutinins”, “Anti-D antibodies”, “Anticomplementary activity”, and “Prekallikrein activator” are quality indicators that allow the tolerance of patients to the administration of these products to be predicated. Methods of evaluation are based on biological effects (agglutination and erythrocyte haemolysis) and the amidolytic properties of product cascade reaction (splitting of specific substrate). These methods are associated with the use of biological origin reagents (complement haemolytic serum, erythrocytes, antiglobulin serum, prekallikrein) and cannot be standardized with respect to all parameters. However, the contemporary practice of drug testing imposes more stringent requirements for the methods and techniques used in the evaluation of their quality. Of fundamental importance in improving the procedures of HIG and HA in terms of security-specific indicators is the use of RMs in testing products, some of which are used to assess the stability of analytical work (the system suitability assessment), while others, together with the test method, are a means of transmitting units of magnitude, allowing the quantitative content of impurities to be established [1].

2 Literature Review

The international practice of making and using RMs to control the quality of immunoglobulin and human albumin products demonstrates a significant variability in methodological approaches. In order to assess the limitation or exceeding of the limiting content of the HAE and PKA of anti-D antibodies, as well as to obtain the level of ACA in the range of more than 50%, RMs are needed that do not currently meet the requirements of the PSF for products according to these quality indicators. However, Ph.Eur. and International (NIBSC) RMs comprise a series of human immunoglobulin or human albumin products released without violations of the technological process. To achieve the corresponding certified value, the RM producer further performs additional “material processing” (also called “manufacturing” or “preparation”) [2] or selection conditions for its use. Thus, human immunoglobulin for anticomplementary BRP activity is a lyophilic-dried, highly purified HIG [3]. An increase in the amount of the initially reconstituted RM solution is a condition for obtaining values in the range of more than 50% (positive control). Depending on the RM series, this amount can be either 0.6 or 0.8 mL when using 0.2 mL for negative control. The level of ACA does not increase when increasing the content of conformationally modified immunoglobulin protein molecules that comprise the trigger mechanism for the cascade of the complement system activation in the patient’s body during immunoglobulin therapy. This level is increased by augmenting the amount of immunoglobulin protein in the CFT technique used to assess the quality of the product by ACA level. For a negative control, any HIG series that meets the requirements of the PSF for ACA can be used. The certified AKA values of this RM appear across in a wide range and, regardless of the series, account for 10–40% and 60–100% for negative and positive controls, respectively. The amount of anti-A and anti-B HAE in immunoglobulin products can be detected by direct and indirect haemagglutination test [4]. Their titre is defined as the maximum dilution of the drug at which the agglutination of test erythrocytes of any intensity degree occurs. In order to control the stability of analytical work across different HAE content determination value ranges, it is necessary to use both positive and negative RMs. To assess the quantitative content, it is necessary to use an RM with HAE content in the titre of 1:64 (RM limit of the HAE content) equal to the maximum allowable for the studied products. Since the content of anti-A and anti-B HAE in donor blood varies, the HAE in the HIG content depends on the prevalence of donors of one or another blood group in the blood plasma production pool. Given this circumstance in the production of positive RMs (International or Ph.Eur.), any amount of anti-A and anti-B HAE in the range from 1:16 to 1:32, which may be either the same or different for both types of HAE, is accepted as a target value. Thus, positive control in the international RM composition (NIBSC) represents a solution of HIG having a different HAE content: anti-A in the range from 1:32 to 1:64 and anti-B in the range from 1:16 to 1:32 [5]. The Ph.Eur. RMs, also comprising a HIG solution, contain both anti-A and anti-B haemagglutinins at the level of 1:32 [3]. The HIG series that does not contain HAE is used for negative RM production. The RMs of the limit of the HAE content (International and Ph.Eur.) represent HIG solutions that have additional murine monoclonal antibodies to achieve the target value of 1:64. These RMs are certified and recommended for use only in the DHA, since murine antibodies cannot be detected in the IHA. The indicated RMs are certified according to the same method in which their use is intended. The content of anti-D antibodies in products of HIG, as determined in the haemagglutination reaction, should not exceed the titre of a positive RM. The anti-D antibody titre is defined as the maximum dilution of the product, at which the agglutination of Rh-positive red blood cells of any degree of intensity occurs. The specificity of the analysis is confirmed using the negative RM, as well as Rh-negative red blood cells. Currently both International and Ph.Eur. RMs are used in general practice [3, 5]. The positive RM, which has a nominal titre of 1:8, is obtained by means of a 6000-fold dilution of the international RM of anti-Rh immunoglobulin with an activity of 285 IU/mL in a HIG solution. The anti-rhesus activity of the positive anti-D antibodies RM content is a calculated value, with the certification for the nominal titre being carried out via the recommended procedure. The assessment of the limiting content of the PKA is carried out using the chromogenic method by comparing the RM with the RM calibrated in the IU [4]. Modern RMs represent a series of HA drugs having a high PKA content, approaching the maximum allowable in the HIG and HA products, which can be achieved by changing certain stages of the product manufacturing process (international RM and Ph.Eur. RM) or by addition of high purity PKA (USP RM) [3, 5, 6]. In order to standardise methods for the assessment of specific safety HIG and HA drugs, the XIII edition was developed and incorporated for the State Pharmacopoeia of the Russian Federation—"Determination of anticomplementary activity of human immunoglobulins", "Determination of anti-A and anti-B haemagglutinins in human immunoglobulins", and the "Test for anti-D antibodies in human immunoglobulins"—according to which standards the use of RMs is necessary for assessing the quality of drugs [7]. Limiting routine analysis methods to international RMs is not economically feasible and fails to ensure compliance with national security principles. In addition, differences in pharmacopoeial analysis methods for HIG products, for example, concerning HAE content, prohibit the use of international RMs in the OFS.1.8.2.0004.15 methods. In accordance with WHO recommendations, national RMs should be developed [8]. For their development, certification and adequate application it is necessary to substantiate the relevant methodological approaches.

3 Materials and Methods

Human immunoglobulin solutions with normalised haemagglutinin content, anti-D antibodies, or anti-complementary activity in a different range were selected as objects of study. In order to determine the quantitative characteristics of the specific safety indicators, complement fixation reaction, indirect and direct haemagglutination, as well as chromogenic and spectrophotometric methods were used in accordance with the SPh RF and Ph.Eur. System analysis and document audit techniques were used to justify the methodological approaches.

4 Results

The method of determining the ACA in accordance with OFS.1.8.2.0007.15 supports the use of RM, the positive and negative controls of which allow us to assess the stability of the analytical work across different ranges of test results. In the development of domestic RM for ACA, the principle of maximum approximation to the impurity content in its composition, which characterise the ACA level, was used. It was determined that immunoglobulin molecules acquire the ability to spontaneously activate the complement system due to aggregation in blood plasma fractionation process. Similar changes in the structure and intermolecular interactions of IgG can be caused by heating, processing with organic solvents and chemical reagents [9]. A method was developed for obtaining a reference material, including the preparation of HIG solutions with an established ACA: (1) for positive control by heating a human immunoglobulin solution to a temperature of 56 ± 1 °C until the ACA level reaches 60–90%; (2) for the negative control, a commercial HIG drug solution having no physical effects if the ACA level is 10–45%. This research allowed the composition and method of using national RM [10] to be justified. National human immunoglobulin RM (IRM 42-28-430) consists of two components: (1) the negative control comprises a HIG solution having an average ACA value of 40.5%, 41.6% and 38.7% (for three certified series respectively); (2) the positive control consists of a heat-treated HIG solution with average ACA values of 76.6, 75.7 and 69.9% [11–13]. The estimated RM uncertainty of these series was about 10%, which indicates a higher stability of the certified characteristics compared to the BRP RM, for which the uncertainty ranges from 15 to 33% [14].

Since the content determination of the anti-A and anti-B haemagglutinins in human immunoglobulin products by indirect haemagglutination is provided by the Russian state quality standard (OFS.1.8.2.0005.15), it is not possible to use the International and the Ph.Eur. RMs for this purpose. In relation to complications of HIG infusion therapy products, the most significant are HAE, which belong to the IgG. These antibodies are only present in the blood of donors with I (0) blood groups. Donor blood II (A) and II (B) contains anti-B or anti-A haemagglutinins (respectively), belonging to immunoglobulins of class M, which are excluded from the composition of the product during the process of fractionation. However, the blood collected from donors having the IV (AB) blood group contains neither anti-A, not anti-B HAE. The use of the appropriate blood group donors to produce plasma candidates for RM in combination with a controlled process for concentrating the final product in accordance with the initial HAE content allowed the RM "Reagent kit for determination of anti-A and anti-B haemagglutinins" (IRM 42-28-439) to be developed; this reagent kit comprises five components [15, 16]. Four of these components (positive) are derived from blood plasma collected from donors with I(0) blood type, but with varying degrees of concentration to achieve the target content value of anti-A and anti-B HAE in the range from 1:16 to 1:64 (for these positive components) and 1:64 for the content limit components of the relevant HAE. The fifth (negative) component does not contain HAE due to the use of blood plasma collected from donors with the VI (AB) blood type. The development of an RM for the determination of HAE content based on HIG solutions containing HAE impurities that cause the occurrence of adverse events of immunoglobulin therapy allows various methods (e.g. DHA and IHA tests) that are regulated to control the quality of HIG products to be used for the certification of the RM. The presence of RMs, whose certified characteristics are established using DHA and IHA tests, not only provides for the standardisation of these methods, but also allows for a comparative assessment of the HAE quantitative characteristics of impurities in various HIG preparations.

The limit of anti-D antibody content in HIG products was established as a result of a retrospective analysis of complications that occur in Rh-positive patients when using products that agglutinate tested red blood cells at a dilution of 1:16–1:32 [17, 18]. When evaluating the content of anti-D antibodies, the dilution of intramuscular anti-rhesus immunoglobulin Partobulin (Immuno Ltd, Sevenoaks, Kent, UK) to a theoretical content of 0.05 IU/mL with a nominal titre of 1:8 was used as a positive control; this served as the basis for a consideration that a positive control in assessing the quality of HIG products should have the indicated activity. The preparation strategy of the International and the Ph.Eur. RMs do not accurately characterise the anti-Rhesus activity of positive RM, since the resulting haemagglutination reaction variability is high, including through the use of two-step dilution. A 6000-fold dilution of certified anti-rhesus HIG also contributes to the uncertainty of anti-D antibody content. It would be more reasonable to use the HIG solution for the development of the RM, obtained from the corresponding rhesus accessories blood plasma of donors with varying degrees of concentration, to achieve the target value of anti-D antibodies content. The flow cytometry method allows the values of the certified content of anti-D antibodies to be set in the IU, while the nominal titre can be determined using the haemagglutination method. The national RM "Reagent kit for determination of anti-D antibodies" (IRM 42-28-440-2017) comprises two components [19]. The negative component made from blood plasma collected from Rh-positive donors having the I (0) blood type does not contain any anti-D antibodies. The target value of the anti-D antibody content in the positive component is achieved using blood plasma collected from Rh-negative donors having the IV (AB) blood group. The positive component is certified by two methods: haemagglutination (titre is 1:8) and by method of flow cytofluorometry (anti-Rh activity is 0.7 IU/mL). The presence of a characteristic certified in comparison with international RM by two methods, one of which (flow cytofluorometry) is more accurate and reliable, allows the development of a positive RM component to be standardised.

Considering the international experience of PKA RM development, the authors of the present study analysed the PKA content in HA commercial batches [20, 21]. It has been established that the technological features of domestic production of drugs from human blood plasma do not permit the use of commercial HA batches as a candidate for RM. Due to the impracticality of changing the HA production stages, it is necessary to consider the possibility of developing a national RM using the highly purified β-fragment of the activated Hagemann factor, which is a genuine PKA contaminant of HA products.

5 Discussion and Conclusions

Thus, RMs play a key role in assessing the specific safety of both HIG- and HA-based products. The development of RMs used to quantitatively characterise impurities, which cause the occurrence of undesirable reactions when using products of immunoglobulins and human albumin, requires the selection of a candidate RM. The national RMs are developed considering the closest approximation in their composition to the corresponding impurity content, the quantitative characteristic of which consists in the certified value. Due to the lack of currently alternative methods for measuring the values characterising the indicators of specific safety, the IRM of human immunoglobulin and the IRM for determining the content of anti-A and anti-B haemagglutinins are both certified according to the same method in which their use is intended. The main advantage of the positive component of the IRM for determination of anti-D antibody content consists in its certification according to two distinct methods.

Acknowledgements The authors sincerely thank the reviewers and the Editorial board for the highly professional and benevolent attitude during the preparation of materials for publication.

References

1. Volkova RA et al (2016) Topical issues related to reference standards in the sphere of circulation of biological products. BIOpreparations Preven Diagn Treat 16(4):229–236. https://doi.org/10.30895/2221-996x-2016-16-4-229-236 (In Russ)
2. GOST ISO Guide 34-2014 (2016) General requirements for the competence of reference material producers. Standartinform Publication, Moscow (In Russ)
3. Ph. Eur. Reference standards. Available via https://www.edqm.eu
4. European Pharmacopoeia EDQM, 9th edn. Subscription required. Available via http://online6.edqm.eu/ep805/#
5. National Institute for Biological Standards and Control. Biological reference materials. Available via https://nibsc.org/products.aspx
6. Food and Drug Administration [site]. Available via https://www.fda.gov/default.htm
7. State Pharmacopoeia of the Russian Federation, XIII edn. Available via http://pharmacopoeia.ru/russian-pharmacopoeia
8. WHO Technical Report Series 897 (2000). Geneva: WHO expert committee on biological standardization. Available via https://apps.who.int/iris/bitstream/handle/10665/42340/WHO_TRS_897.pdf?sequence=1
9. Borg AL (2009) Investigation of method for determination of anticomplementary activity (ACA) in Octagam®: master's thesis. Stockholm. Available via http://liu.diva-portal.org/smash/get/diva2:241409/fulltext01.pdf
10. Krivykh MA, Kornilova OG, Kudasheva EY (2015) A method for obtaining a positive control of an RM of human immunoglobulin to determine the anti-complementary activity of human immunoglobulin preparations, and an RM of human immunoglobulin to determine the anti-complementary activity of human immunoglobulin preparation. RF Patent, 2577703 C1, 9 Feb 2015 (In Russ)
11. Krivykh MA et al (2015) Development of a reference standard for the determination of anticomplementary activity of human immunoglobulin. Khimiko-Farmatsevticheskii Zhurnal 49(6):40–42. https://doi.org/10.30906/0023-1134-2015-49-6-40-42 (In Russ)
12. Krivykh MA et al (2016) Certification of standard samples of human immunoglobulin for determination of anticomplementary activity. Khimiko-Farmatsevticheskii Zhurnal 50(12):61–64. https://doi.org/10.30906/0023-1134-2016-50-12-61-64 (In Russ)
13. Kornilova OG et al (2017) Contemplation of the possibility of extending the shelf life of human immunoglobulin reference standard used for determination of anticomplementary activity. BIOpreparations Prev Diagn Treat 17(2):110–115. https://doi.org/10.30895/2221-996x-2017-17-2-110-115 (In Russ)
14. Sandberg E, Daas A, Behr-Gross M-E (2006) Collaborative study to establish human immunoglobulin BRP Batch 3 and human immunoglobulin (molecular size) BRP batch 1. Pharmeuropa Bio 1:37–49
15. Kornilova OG, Kudasheva EIu, Nechaev AV, Borisevich IV (2017) Reference material of anti-A and anti-B haemagglutination content in human blood preparations. RF Patent, 2017119536 (In Russ)
16. Kornilova OG, Nechaev AV, Borisevich IV, Kudasheva EYu (2017) Standardization of the method for determining haemagglutination content in human immunoglobulin preparations in Russia. Khim-Farm Zh 55(11):57–60. https://doi.org/10.30906/0023-1134-2017-51-11-57-60
17. Pisani G, Wirz M, Gentili G (1996) Anti-D testing in intravenous immunoglobulins: shouldn't it be considered? Vox Sang 71:132
18. Turner CE, Thorpe S, Brasher M (1999) Anti–Rh D activity of commercial intravenous immunoglobulin preparations. Vox Sang 76:55–58

19. Kornilova OG, Kudasheva EIu, Nechaev AV, Struchkova IN, Borisevich IV (2017) Reference material of anti-D antibody content in human immunoglobulin preparations. RF Patent application № 2017144239 (In Russ)
20. Krivykh MA, Kornilova OG, Zhuk IE (2017) Prekallikrein activator content in domestically-produced human albumin products. Medical Immunology 19(S):271–272 (In Russ)
21. Tomilin MV, Filatova EV, Kuznetsova MM, Sudakova VV, Zubkova NV (2017) Validation of the method for determination of prekallikrein activator in Human Albumin. BIOpreparations Prev Diagn Treat 17(1):59–64. https://doi.org/10.30895/2221-996x-2017-17-1-59-64 (In Russ)

Reference Materials for Ensuring the Metrological Traceability of Enzyme Catalytic Activity Measurement (KATAL)

Elena V. Kulyabina, Olga N. Melkova, Ekaterina A. Guskova and Tatiana V. Grebennikova

Abstract The article considers problems associated with the metrological support of catalytic activity measurement. The paper presents the characteristics of the State Primary Special Measurement Standard for the unit of catalytic activity—KATAL—and examines the role of reference materials (RMs) in ensuring the metrological traceability of measurement results. A method for measuring the concentration of the recombinant protective antigen of Bacillus anthracis and the recombinant GP-protein of the Ebola virus, involving the use of a "sandwich"-type enzyme-linked immunosorbent assay, was certified by VNIIMS as a measurement procedure. The following RMs tested by VNIIMS can be used for the development and production of appropriate control samples when performing measurements for comparisons: the RM for the mass concentration of the recombinant protective antigen of Bacillus anthracis in phosphate-saline solution, the RM for the mass concentration of the recombinant GP-protein of the Ebola virus and the RM for the mass concentration of recombinant Clostridium Difficile toxin. The paper presents the main reactions of the method for measuring the catalytic activity of catalysts for heterogeneous processes, which are also used to carry out reaction gas cleaning processes. Thus, conditions have been created for building hierarchies of calibrations of in-demand objects established on the basis of the List of Critical Technologies of the Russian Federation and state programs for the development of industrial sectors.

Keywords Metrological traceability · State primary measurement standard KATAL · Catalytic activity · Catalytic concentration · Laboratory medicine · Reference materials of biological media · Heterogeneous solid-state catalysts · Calibration hierarchies

---

E. V. Kulyabina (✉) · O. N. Melkova · E. A. Guskova
All-Russian Research Institute of Metrological Service (VNIIMS), Moscow, Russia
e-mail: kuliabina@vniims.ru

T. V. Grebennikova (✉)
N. F. Gamaleya Federal Research Center for Epidemiology and Microbiology of the Ministry of Health of the Russian Federation (NITSEM Named After N. F. Gamaleya of the Ministry of Health of the Russian Federation), Moscow, Russia
e-mail: t_grebennikova@mail.ru

S. Medvedevskikh et al. (eds.), *Reference Materials in Measurement and Technology*,
https://doi.org/10.1007/978-3-030-32534-3_6

Abbreviations

ELISA	Enzyme-linked immunosorbent assay
IRM	Flow of the initial reaction mixture
FRM	Flow of the final (contact) reaction mixture
CFR	Circulating flow reactor

1 Introduction

Every human desires to be healthy, wealthy and to live a long and happy life. In order to support this aspiration, it is important for every person to have reliable information about the status of his or her organism, including test results, diagnosis and administration of proper treatment. This necessitates a robust social healthcare system supported by the existence and full operation of a metrological system in laboratory medicine for which metrological traceability of measurement results is a prerequisite.

In accordance with Federal Law of 26 June 2008 No. 102-FZ "On Ensuring the Uniformity of Measurements", [1] the notion of "traceability" is a property of the standard measurement unit, instrument or result, which consists in documenting the establishment of their connection with the state primary measurement standard or the national primary measurement standard of a foreign state of the corresponding unit for measurement standards and their verification, as well as the calibration of measuring instruments.

Metrological traceability [2] is a property of the measurement result, with the help of which the result can be correlated with the basis for comparison through documented continuous calibration chain, each link of which contributes to the measurement uncertainty.

We see our role in providing conditions for improving the quality of the obtained test results, as well as in providing comprehensive metrological support to clinical diagnostic and chemical toxicology facilities as well as those that measure the parameters of the human body.

The goal is achieved by: (1) development of the State Primary Special Standard of the unit of catalytic activity (KATAL); (2) participation in international comparisons; (3) the development of substance reference materials (RMs) under analysis in serum and blood plasma; (4) the development and certification of measurement methods including reference measurement methods; (5) the creation of digital metrology for laboratory medicine.

The metrological characteristics of the experimental sample of the State Primary Measurement Standard for the unit of catalytic activity of biological and chemical substances are given in Table 1 [3].

The important question of developing RMs for supporting the measurement of catalytic activity must correspond to measuring tasks and be commutable between

Table 1 Metrological characteristics of the State Primary Special Measurement Standard for the unit of catalytic activity of biological and chemical substances (KATAL)

Name of metrological and technical characteristics	Value of metrological and technical specifications
Measurement range of catalytic activity (catalytic concentration) of biological substances (kat/dm^3)	From 7.8×10^{-7} to 9.4×10^{-6}
Measurement range of catalytic activity (specific catalytic activity) of chemicals (kat/g)	From 1×10^{-8} to 2×10^{-7}
Expanded uncertainty of catalytic activity measurements (k = 2) (%)	From 0.5 to 4

analytes. Manufacturers and developers of modern methods and diagnostic tools require metrological support for developing the testing, verification and calibration of new measuring instruments, as well as appraisal of the developed techniques. For such measuring instruments, metrological traceability must be ensured by using solutions having known catalytic activity (catalytic concentration) and a known concentration of target analytes approved as RMs.

Although foreign manufacturers produce calibration mixtures for factory testing of their analysers, which can potentially be used for metrological provision of modern equipment, they are expensive and their composition, as a rule, is subject to commercial secrecy, hampering their approval as RMs in the Russian Federation. Our biological and IT metrological support laboratory makes a considerable contribution to solving this problem. In 2018, we began the development and testing of RMs for the catalytic activity of a number of enzymes of the classes of oxidoreductases and hydrolases intended for the transfer of units of catalytic activity, calibration and verification of bioanalytical measuring instruments.

Together with NITSEM named after N. F. Gamaleya of the Ministry of Health of the Russian Federation in 2017, we also approved three RMs—mass concentration of recombinant protective antigen Bacillus anthracis in phosphate-saline solution, mass concentration of recombinant GP protein of Ebola virus and mass concentration of recombinant toxin Clostridium Difficile—whose metrological characteristics are given in Table 2.

2 Materials and Methods

2.1 RMs of Biological Media

Tests of mass concentration RM of recombinant protective antigen Bacillus anthracis and the mass concentration of recombinant GP protein of the Ebola virus were tested in the framework of metrological support prototypes of Monoclon-Bio test-systems

Table 2 Metrological characteristics of certified reference materials of biological media tested by our laboratory in 2017[1]

Name of certified characteristic	Range of permissible certified values (ng/mL)	Boundaries of permissible values of the relative error δ, at P = 0.95 (%)
GSO 10922–2017 Mass concentration of recombinant protective antigen of Bacillus anthracis	From 20 to 30	±22
GSO 10921–2017 Mass concentration of recombinant GP protein of the Ebola virus	From 20 to 30	±22
GSO 10920–2017 Mass concentration of recombinant toxin Clostridium difficile	From 500 to 1000 (ng/μL)	±7

[1]GSO 10920–2017 RM of the mass concentration of a recombinant toxin clostridium difficile in the physiological buffer. Available via https://fgis.gost.ru/fundmetrology/registry/19/items/389281
GSO 10921–2017 RM of mass concentration of recombinant protein gp of ebola virus in phosphate-salt solution. Available via https://fgis.gost.ru/fundmetrology/registry/19/items/389280
GSO 10922–2017 RM of mass concentration of recombinant protective antigen bacillus anthracis in phosphate-salt solution. Available via https://fgis.gost.ru/fundmetrology/registry/19/items/389279

designed to indicate and identify pathogenic biological agents based on a monoclonal antibodies panel.

The system is based on the method of determining low concentrations of specific antigens using the "sandwich" variant of enzyme-linked immunosorbent assay (ELISA), including three main steps: the formation of the immune complex "antigen (test substance)—a specific antibody" or vice versa; the formation of a conjugate connection with the immune complex formed at the previous stage or with free binding sites (determinants); the transformation of the substrate under the action of the enzyme label in the recorded signal as a result of the biochemical reaction.

The test system composition includes 20 individual packages with specific components for each of the desired pathogenic biological agents and non-specific components. The system uses an approved type of flatbed vertical scanning spectrophotometer with a 450 nm light filter and photometers for imark microplates manufactured by Bio-Rad (USA). To ensure the metrological traceability of this system, a measurement procedure was developed and certified, which determined the relationship between the measured optical density and the analyte concentration. Measurement techniques were developed, manufactured and tested for approbation and accuracy control. This was done to confirm the type of RM mass concentration of recombinant protective antigen Bacillus anthracis in phosphate-saline solution as well as the mass concentration of recombinant GP Ebola virus protein. The photometric measurement method was used. The range of concentration measurements ranged from 20 to 30 ng/ml.

In order to control the accuracy of the method, a "zero" sample (i.e., one that does not contain protein) was measured prior to determining the concentration of the drug. The concentrations of the recombinant protective antigen Bacillus anthracis and the recombinant GP protein of the Ebola virus in the corresponding control samples were measured using the components of the Monoclon-Bio test system. The analysis duration was 3.5–4 h. Immediately after stopping the reaction, the optical density was measured in each hole at 450 nm (A_{450}) using the spectrophotometer.

3 Research Results

According to the titration data, a calibration curve was constructed, where the value of the protein concentration of RM was plotted along the X-axis of the abscissa, while the value of the optical density at 450 nm was plotted along the ordinate axis (Fig. 1).

The constructed calibration curve was used for processing the measurement results. The A_{450} signal direction, in the range of 1000–1500, corresponded to the protein concentration of the recombinant protective antigen of Bacillus anthracis or the recombinant GP protein of the Ebola virus in positive control samples (K+) within 20.0–30.0 ng/ml respectively. This technique and corresponding RM can be used for the development and production of relevant control samples, carrying out measurements for key and pilot comparisons, in interlaboratory comparison tests, as well as the for calibration and verification of measuring instruments for analytical purposes.

Our programme also includes the testing of RMs of catalytic activity of catalysts for heterogeneous processes, which are used for purposes including the implementation of exhaust gas cleaning processes.

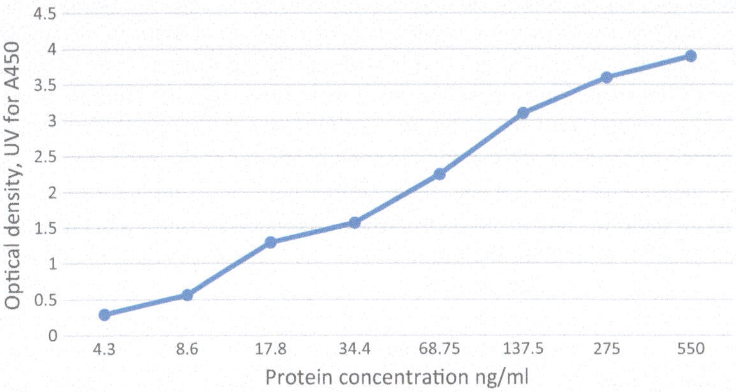

Fig. 1 Calibration curve reflecting dependence of A_{450} on protein concentration in the CRM

The quality of solid-state catalysts is primarily determined by their activity, which, in turn, depends on a complex of various indicators, such as component concentration, surface area, size of catalyst granules, etc.

In 2018, tests were carried out on the alumina-chromium catalyst IK-12-72 and the oxide-alumina-manganese catalyst ICT-12-40 RMs, developed by Boreskov Institute of Catalysis, Russian Academy of Sciences (Russia), where the catalytic activity was expressed in specific units of catalytic activity (katals), referred to as grams of catalyst.

3.1 Aluminum Magnesium Chromium Catalyst IK-12-72

The aluminum magnesium chromium catalyst IK-12-72 is represented by a ball of dark green colour with a diameter of 1.0–1.6 mm. The catalyst consists of a Cr_2O_3, MgO and Al_2O_3 alloy. The range of catalytic activity (specific catalytic activity) of the IK-12-72 alumina-chromium catalyst is from 8.8×10^{-8} to 9.5×10^{-8} kat/g. The absolute error ($k = 2$) is given by 1.5×10^{-8} kat/g.

3.2 Alumina Manganese Oxide Catalyst ICT-12-40

The alumina manganese oxide catalyst ICT-12-40 comprises a cylinder with a diameter of 3 mm and a length of 3–5 mm. This catalyst combines Mn_2O_3 and Al_2O_3. The range of catalytic activity (specific catalytic activity) of the ICT-12-40 alumina-manganese oxide catalyst is from 3.8×10^{-8} to 5.3×10^{-8} kat/g. The absolute error ($k = 2$) is 0.7×10^{-8} kat/g.

Specific catalytic activity is a normalized metrological characteristic of RMs of heterogeneous processes. The method for measuring the specific catalytic activity of these catalysts involves the measurement of catalytic activity in the reaction of complete oxidation of carbon monoxide with molecular oxygen in air at a constant temperature. The following measuring instruments and auxiliary equipment are used for the implementation of the technique: a SOLO catalytic installation (Russia), a gas chromatograph, an analytical balance and a laboratory thermometer.

When measurements are made, three source gases are supplied to the catalytic unit using air metering units—oxygen, carbon monoxide and nitrogen. Flows of these gases are created, which are then mixed with the formation of a flow of the initial reaction mixture (IRM); the gas flow rate changes according to the instructions of the operator during the experiments. The flow of the initial reaction mixture passes through a circulating flow reactor (CFR). The main parts of the CFR unit comprise a thermostated flow tube reactor and a circulating pump. The test sample is placed inside the reactor in the form of a granular catalyst layer. Upon contact of the reaction mixture with the surface of the heated catalyst, a proportion of the starting materials is converted into reaction products according to Eq. (1) and the flow of the final (contact) reaction mixture (FRM) leaves the CFR unit.

The exothermic reaction of complete oxidation of carbon monoxide, selected as the key reaction component, is described by Eq. (1).

$$2CO + O_2 \rightarrow 2CO_2 + 566\frac{kJ}{mol} \quad (1)$$

This reaction proceeds without changing the volume of the reaction mixture, making it possible to investigate the dependences of the reaction rate on the concentrations of two starting materials—carbon oxide and oxygen, as well as two reaction products—carbon dioxide and water.

The specific stationary reaction rate of the complete oxidation of carbon monoxide (R) is calculated by Eq. (2).

$$R = \frac{(C^{\circ} - C) \cdot U}{m} = \frac{C^{\circ} \cdot X \cdot U}{m} \, cm^3/s \cdot g, \quad (2)$$

where C° and C—concentrations of carbon monoxide in the IRM and FRM, respectively, mass fraction (ppm); m—catalyst mass, g; U—the rate of supply of the IRM; X—the degree of carbon monoxide conversion, calculated by Eq. (3).

$$X = \frac{C^{\circ} - C}{C} 100\% \quad (3)$$

The carbon monoxide concentrations in the IRM and FRM are determined by gas chromatograph with a thermocatalytic detector and a chromatographic column based on activated carbon (SKT, Russia). The analysis time is 60 s. The oxygen and carbon dioxide concentrations in FRM are calculated according to the formulas (4) and (5).

$$C_{O_2} = 2C^{\circ}_{O_2} - \left(C^{\circ}_{CO} - C_{CO}\right) \text{ mass fraction} \quad (4)$$

$$C_{CO_2} = C^{\circ}_{CO_2} + \left(C^{\circ}_{CO} - C_{CO}\right), \text{ mass fraction} \quad (5)$$

where $C^{\circ}_{O_2}$ and $C^{\circ}_{CO_2}$—the oxygen and carbon dioxide concentrations in the IRM, respectively, *mass fraction.*

To recalculate the specific catalytic activity value according to Eq. (2) from ml/g s or cm^3/g s to kat/g, the following expression can be used (6).

$$\frac{mkat}{g} = \frac{mmol}{s\,g} = \frac{ml}{22.4\,sg} \quad (6)$$

The corresponding values of the stationary reaction rates are determined by alternately specifying the rate of the IRM supply to the reactor. A graph of the stationary

Table 3 Conditions for measuring the activity of catalysts IK-12-72 and ICT-12-40

Parameter	Value
Catalyst sample intake (g)	2.5
Test temperature	250
IRM composition	CO—5.0 vol%, O_2—5.0 vol%, the rest is purified nitrogen
Volumetric IRM feed rate (ml/min)	100–1000

reaction rate R on the concentration of carbon monoxide in FRM on a logarithmic scale (kinetic graph) is constructed using the obtained data. From the obtained graph, the value of the stationary reaction rate (R) at a given level of conversion of carbon monoxide to FRM (C = 50%), which is the value of catalytic activity, is either determined analytically or graphically.

Taking into account the developed methodology for measuring specific catalytic activity, a series of measurements of the activity of the IK-12-72 and ICT-12-40 catalysts were carried out. Conditions for measuring the activity of catalysts are given in Table 3.

4 Research Results

As a result of the experiments, the specific catalytic activity values of IK-12-72 and ICT-12-40 were obtained, which are listed in Table 4. The accuracy of the obtained results was evaluated in terms of uncertainty. According to the conducted experiment results, it was revealed that the greatest contribution to overall measurement uncertainty is made by the measurement uncertainties of the key component concentration and the IRM flow rate. These contributions exceed all others by an order of magnitude. It is planned to identify the main reasons for the occurrence of these contributions as well as measures for reducing them in a future study.

The homogeneity and stability studies for the experimental RM samples are currently ongoing.

Table 4 Results of the experimental determination of the specific catalytic activity of catalysts IK-12-72 and ICT-12-40

№	Catalyst	Specific activity (when conversion level C = 50%) (μkat/g)	Uncertainty (%)
1	ICT-12-40	0.0543	10.5
2	IK-12-72	0.1200	11.4

5 Conclusions

The result of ongoing work on the RM development, providing accurate results of catalytic activity measurements, will be the construction of calibration hierarchies for the sought-after objects established on the basis of the Russian Critical Technologies List and both "Development of the pharmaceutical and medical industry 2013–2020" and "Development of industry and increase its competitiveness for the period up to 2020" programs.

The created experimental sample of the State Primary Measurement Standard Unit of the Catalytic Activity of Biological and Chemical Substances (KATAL) provides metrological traceability for measurements of biologically active components in solutions, liquids and human tissues by means of a chain of calibrations using state-owned RMs of catalytic activity.

Thus, as a result of obtaining reliable results of tests on human organisms, it is possible to foresee the creation of conditions for more accurate diagnoses, the correct prescription of therapeutic treatments and improved quality of life and longevity. There is reason to hope that the implementation of these identified works will bring the dream of a long and healthy life closer.

References

1. Federal Law (2008) On ensuring the uniformity of measurements. No FZ-102 of 26/06/2008. Available at http://www.fundmetrology.ru/depository/01_npa/102-fz_2015.pdf (In Russ)
2. International Vocabulary of Metrology (2010) Basic and general concepts and associated terms. Translation from English and French is made. D. I. Mendeleyev Institute for Metrology, Belarusian State Institute of Metrology, 82 p (In Russ)
3. The final report on the implementation of scientific research work. Conducting fundamental research in the field of measurements of the physicochemical composition and properties of substances with the aim of creating a State Primary Special Measurement Standard for the unit of catalytic activity of biological and chemical substances—KATAL. VNIIMS 1:4 (2016)

Manufacturing and Certification of Reference Materials for the Mass Fraction of Boron in Silumin

Dmitry G. Lisienko, Margarita A. Dombrovskaya and Eugeniya D. Kubrina

Abstract In this research, we set out to provide metrological support to methods used for determination of the content of boron in silumin. In particular, we present a technology for the synthesis and characterisation of reference materials for the composition of silumin. To this end, a number of research methods were applied, including X-ray phase analysis, atomic emission spectral analysis with arc excitation, gravimetry and mass spectral analysis. As a result, the developed set of reference materials has been included into the State register and the Federal information fund.

Keywords Set of CRMs for the mass fraction of boron in silumin · Synthesis · Estimation of metrological characteristics · Homogeneity · Stability · Traceability

1 Introduction

Aluminium alloys are widely used for various industrial purposes. When applied in the nuclear power industry, these alloys may present danger due to such their critical admixture as boron. Boron is known to possess a high thermal neutron absorption capacity with the effective capture cross section of 752 barn. Therefore, the boron content limit in aluminium alloys, particularly in silumin and electrocorundum, is typically set at a fairly low level, i.e. its mass fraction should not exceed the value of $3 \times 10^{-3}\%$.

The determination of the mass fraction of boron for its contents over $1 \times 10^{-4}\%$ is carried out by a highly sensitive atomic-emission method with arc excitation after the preliminary conversion of the studied object into an oxide form. The analytical testing centre "Russian Arbitration Laboratory for Testing Nuclear Energy Materials", Ural Federal University, following the order of the "Mashinostroitelny Zavod", JSC, developed and tested a set of reference materials (RMs) for the mass fraction of boron in silumin (SOBor). These RMs are intended for the calibration of equipment used in the atomic emission analysis of silumin and electrocorundum, certification

D. G. Lisienko · M. A. Dombrovskaya (✉) · E. D. Kubrina
Ural Federal University (UrFU), Ekaterinburg, Russia
e-mail: margodomb@yandex.ru

S. Medvedevskikh et al. (eds.), *Reference Materials in Measurement and Technology*,
https://doi.org/10.1007/978-3-030-32534-3_7

Table 1 Intervals of the mass fraction of boron (%) in the RMs

RM type index			
SOBor-1	SOBor-2	SOBor-3	SOBor-4
$(2.5–3.5) \times 10^{-3}$	$(0.5–1.5) \times 10^{-3}$	$(1.6–4) \times 10^{-4}$	$(0.7–1.5) \times 10^{-4}$

and control of the accuracy of methods for measuring the composition of silumin (electrocorundum).

In the technical specification, the Customer formulated the following basic requirements that the developed RMs should meet:

- fabrication of a set of 4 RM types, covering a large range of boron concentrations and 1.5 orders of magnitude (Table 1);
- high stability ensuring a stable functioning of the RMs for several decades;
- the error of the certified values not exceeding ½ of the error of the analytical method;
- the proximity of the spectrochemical characteristics of the developed RMs to those of the samples under study;
- a relatively small amount of release (within the limits of 0.5 kg for each RM).

The limits of the permissible relative error are lower than 6%.

During the RM fabrication, we implemented a technology that had been repeatedly tested in our previous studies [1, 2]. It consists in the introduction of a metered quantity of the component to be certified into a specially prepared matrix material. This method allows the metrological characteristics of RMs to be established by experimental calculation procedure specified in the State requirements MI 1992-98 [3].

2 Experimental Part, Results and Discussion

2.1 Selection and Preparation of Materials for the Synthesis of RMs

The choice of starting materials and the preparation of an RM matrix presented a certain difficulty. According to the method of silumin analysis [4], an analytical sample should be represented by an oxidised alloy. In order for silumin to be converted to an oxidised state, it is dissolved in a mixture of hydrochloric and nitric acids. The solution is subsequently dried, and the salts are calcined to form stable oxides at 800–900 °C. The preparation of an RM matrix by means of metal oxidation turned out to be impossible due to the lack of silumin with a required purity. In addition, the dissolution of the metal requires the use of acids in large volumes, which may cause the introduction of significant boron contamination into the substrate. Therefore, it was decided to create an artificial matrix. The fact that the analysed metal contained 86% of Al, 12% of Si and 2% of Ni was taken into account. The X-ray analysis of the alloy in the oxidised form, performed using a URD 6 Bruker AXS GmbH diffractometer (Germany) in copper K_α radiation, showed the material to be a mixture of aluminium and nickel oxides, as well as elemental silicon. This composition was considered to be consistent with the literature data on the forms of silicon in aluminium alloys [5]. Thus, for the matrix RM material to be adequate to oxidised silumin in terms of chemical form, it should represent a mixture of the abovementioned compounds in amounts corresponding to the content of the elements in the alloy.

The starting material for the production of aluminium oxides was analytically pure ammonium alum according to GOST 4238-77, which was calcined in alundum crucibles in a muffle furnace at 1000 °C to constant weight. This heat treatment mode permits the most chemically stable form of aluminium oxide—α-modification—to be obtained [6]. The oxide was subsequently crushed and averaged in a cylindrical plexiglass mixer. Silicon powder in the elemental state was obtained from monocrystalline silicon by crushing and grinding in a roller mill. To isolate the fraction of the desired size (-100 μm), the powder was sieved through a sieve fixed in a hermetically sealed two-chamber body. Nickel was introduced into the substrate in the form of a green oxide, which had been obtained by a heat treatment of a reactive high-purity black nickel oxide (10–2) at 850 °C according to GOST 4331-78. All the powders were averaged in the mixer described above.

In order to assess the boron content and confirm the possibility of using the selected substances as matrix components, the resulting substrate was analysed by the atomic emission method with arc excitation during the calibration of a PGS-2 spectrometer (Carl Zeiss Jena, GDR) with a linear solid-state radiation detector of MAES (VMK-Optoelectronics, RF) by the method of additives. The role of "additives" was performed by RMs for the composition of graphite of approved type, a SOG-21 set (CRM GSO 4521-89/4523-89, CRM MSO 1356: 2007, COOMET CRM 0102-RU-2011) [7]. Due to the proximity of the mass fraction of boron in the substrate ($\leq 1 \times 10^{-4}$%) to that contained in SOBor-4 (Table 1), it was decided to use the matrix material as an RM sample with the minimum concentration of the component under certification. In this case, a corresponding characterization of this material should be carried out in order to take into account the boron content in the matrix when calculating the composition of all RMs.

10-aqueous sodium tetraborate was chosen as a carrier for the component under certification, the mass fraction of boron in which was established by certified method [8].

2.2 *Synthesis of RM Samples*

A sample with a maximum boron content (SOBor-1) was obtained by introducing an aqueous solution of sodium tetraborate into the matrix followed by heat treatment at 300 °C for 4 h, grinding and homogenization in a cylindrical plexiglass mixer. A portion of 10-aqueous sodium tetraborate was weighted using an analytical balance AB 104-S/FACT (Mettler-Toledo GmbH, Switzerland) [9] of a special accuracy class according to GOST R 53228-2008. To prepare a solution, deionized water with a specific electrical resistance of 18.2 MΩ cm was used. The aliquot of the solution required to ensure a given concentration of boron in a sample was dosed gravimetrically. The use of this technology allows the following RM indicators to be relatively simply provided: high uniformity of the distribution of the component under study in the matrix; transfer of the material into a stable chemical form during heat treatment, with the fixation of the component under study on the matrix particles, which determines a high segregation stability of RMs. It was experimentally proven that no loss of the component under certification and no uncontrolled contamination of the materials occur during all such synthesis operations. The absence of boron loss during calcination is provided by the solvent type and heat treatment regimes.

Another two RM samples were synthesised by a sequential dilution of SOBor-1 with the matrix and subsequent averaging of the materials. To ensure a more uniform initial distribution of boron by volume, averaged RM portions weighted using a balance GX 1000 (A&D Co., LTD., Japan) were placed in a mixer alternatively in small portions. The calculation of the matrix mass was carried out taking into account the moisture content in the matrix.

The implemented method of RM preparation allows the certification of the synthesized materials to be carried out using a calculation method taking into account data on the boron content in the matrix and the degree of sample inhomogeneity.

2.3 *Testing and Characterization of RMs*

The calculation of the concentration of the component under study in the RMs was based on data about the composition of the reagent that acted as a carrier of the certified component, the composition of the matrix, as well as results of their weighing and weighing of finished materials. When assessing the error of certified values, all its sources were taken into consideration, including errors induced by the determination of the composition of starting materials, fabrication procedures, inhomogeneous distribution of the component under study and sample instability.

The former two error components (composition determination and fabrication procedure) were calculated on the basis of the law of error accumulation, relying on the information about the characteristics of the substances used in the synthesis and measuring instruments (balances). The following components of error were taken into account, including those induced by: determination of the mass fraction of metals in the RM matrix; weighting of the portions of matrix components; establishment of the main substance fraction in a component; measurement of the mass fraction of boron in sodium tetraborate; determination of the mass fraction of boron in the solution for introduction; measurement of the mass of the solution introduced into the matrix.

Since the certified characteristic is the mass fraction of boron, which is calculated relative to the sum of all elements in silumin, excluding oxygen, the hydroscopicity of aluminium oxide makes the main contribution to the error induced by determination of the mass fraction of metals in the matrix. Losses related to calcination amounted to (3.06 ± 0.17)% for alumina.

The analysis of the matrix material is a difficult stage in any certification research. The certified value of the mass fraction of boron was established by the method of additives in the Russian Arbitration Laboratory for Testing Nuclear Energy Materials (UrFU) and the research laboratory of "Mashinostroitelny Zavod" (MSZ) according to the certified method [2] using SOBor-1–3 RMs as samples. When the method was implemented under the conditions of the MSZ laboratory, four results were obtained for four SOBor-4 portions. Similarly, four measurements for 10 SOBor-4 portions were performed at the UrFU laboratory. Co-processing of values (Table 2) with the assessment of the absence of significant differences between their relative dispersions (CV_R^2) according to the Cochran G (within one organisation) and Fisher F tests for the two generalized dispersions of each laboratory (Note to Table 2) allowed us to calculate the average result and the generalised relative standard deviation (GRSD) of reproducibility characterizing (CV_R, %), on the basis of which the limits of the error characteristic of the mass fraction of boron in the substrate were calculated (± Δ) as

Table 2 Determination of boron concentration in SOBor-4 RMs

Laboratory	W (%)	CV_R (%)	f
MSZ	1.110×10^{-4}	1.2	12
	1.214×10^{-4}	8.0	
	1.191×10^{-4}	4.7	
	1.103×10^{-4}	4.9	
Ural Federal University	1.070×10^{-4}	6.2	30
	1.004×10^{-4}	7.9	
	1.085×10^{-4}	1.4	
	9.63×10^{-5}	7.8	
	1.035×10^{-4}	12.5	
	9.70×10^{-5}	13.5	
	9.98×10^{-5}	11.5	
	9.43×10^{-5}	8.7	
	1.045×10^{-4}	15.1	
	9.35×10^{-5}	11.3	
Average value	1.018×10^{-4}	9.2	42

Note $G_{MSZ} = 0.60$ under Gcrit (0.95; 4; 3) = 0.68 [10, p. 243]; $G_{UrFU} = 0.16$ under Gcrit (0.95; 10; 3) = 0.73 [10, p. 243]; $F = 1.10$ under Fcrit (0.95; 30; 12) = 2.47 [10, p. 208]

$$\Delta = \pm t_{0.975;42} \cdot CV_R \cdot W/100,$$

where $t_{0.975;42}$ is the Student t-distribution quantile for a one-sided confidence level of 0.975 and a number of degrees of freedom 42 equal to 2.02 [10, p. 178].

Therefore, the boron content in the matrix material (SOBor-4) is assumed to be $(1.018 + 0.028) \times 10^{-4}\%$.

When planning experimental work to study the degree of RM heterogeneity, we took into account that the SOBor RM preparation technology provides a fairly uniform distribution of the component under certification. This technology allows the amount of experimental research to be optimised. To confirm the validity of these findings and to establish the dependence of the degree of heterogeneity on the concentration of the component under certification, SOBor-1 and SOBor-2 RMs were investigated. Since the distribution of boron in Sobor-4 is dependent on both its total content in all matrix components and the quality of the matrix homogenisation, we also conducted studies for this material. All experiments were based on the recommendations of GOST 8.531 [11], which implies the use of single-factor analysis of variance. To this end, 10 samples weighing about 400 mg each were randomly selected from the materials under study during the unloading of the mixing tank. It is known that the smaller the random measurement error, the higher the reliability of conclusions about the distribution of the component under study. Therefore, for the analysis of SOBor-1 materials, a certified mass spectral method [12] was applied using an ELAN 9000 spectrometer (PerkinElmer, USA). The evaluation of the chemical homogeneity of SOBor-4 was performed by the atomic-emission method with arc excitation, since it is practically impossible to quantitatively transfer boron to a solution from a matrix material without loss. The standard deviation value of the heterogeneity of a SOBor-3 sample was calculated based on the dependence of this indicator on the concentration of the element being certified in logarithmic coordinates (Fig. 1). The characteristics of heterogeneity (CV_h, %) for all RM samples of the smallest representative mass of 0.1 g is displayed in Table 3.

When evaluating the RM stability, the method of accelerated aging [13] was used. Variable factors included thermal factors and, taking into account the fine state of materials, vibration load. Investigations were carried out for a sample of SOBor-1 characterised by the smallest certification error and the possibility of recording an analytical signal reliably.

The experiments implied a comparison of the analytical signals generated under identical conditions by a sample having been kept at a temperature of $T_{exp} = 800$ °C for 3 h and by a thermally untreated sample [13]. The certified values are found to be stable for 6×10^{16} years.

Vibration exposure was carried out on a vibration unit [14] in a mode that provides active separation of mechanically mixed powders under vertical and horizontal positions of containers loaded with the material. The amplitude of the circular oscillations of the table was 2 mm at a frequency of 25 Hz, with the processing time being 15 min. The results of the analysis established negligible differences in the concentration of boron in RM portions selected from different zones of the reservoir, which proves a sufficient resistance of the material to delamination. We believe that the long-term

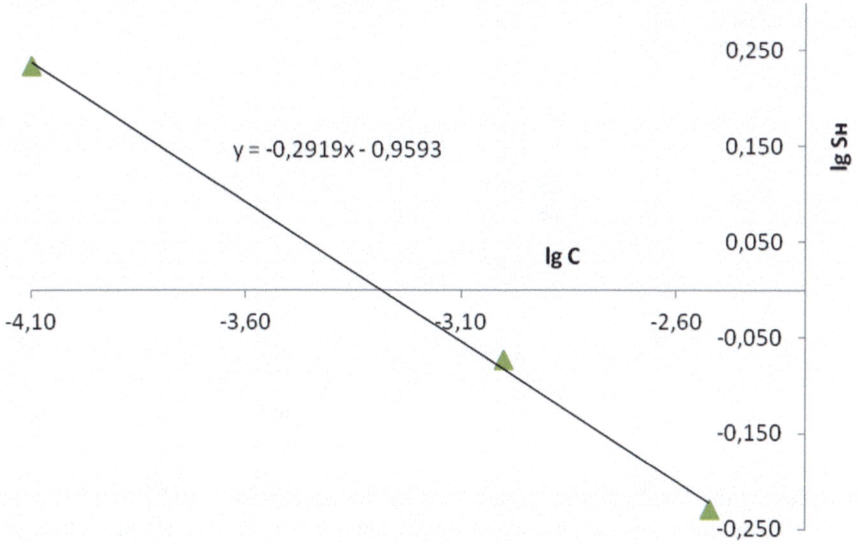

Fig. 1 Dependence of lg S_n on the logarithm of the concentration of the certified component

Table 3 Relative standard deviation of the RM inhomogeneity (CV_h, %)

RM index	RM portion when assessing the indicator (mg)	CV_h for 0.1 g (%)
SOBor-1	10	0.59
SOBor-2	20	0.84
SOBor-3	–	1.09
SOBor-4	40	1.71

stability of the developed RMs can be ensured by packing them in small portions (10 g) in hermetically sealed plastic containers immediately after preparation. The expiry date of the RMs is established to be 60 years, taking into account the time of complete consumption of the materials.

The mutual consistency of the RMs in the set (Fig. 2) was confirmed during the implementation of a procedure, for the metrological support of which the developed RMs are intended [2].

The dependence in the coordinates "lg *I/Ic*–lg*C*", where *I/Ic* is an analytical signal in the form of the relative intensity of the boron line, was drawn based on the results of averaging 14–16 measurements for each RM sample and approximated by the linear equation $y = (0.96 \pm 0.14)\ x + (3.35 \pm 0.46)$. The insignificant difference of the determination coefficient from 1 ($R^2 = 0.994 > r^2 = 0.903$, where $r_{(0.975;2)}$ is the correlation coefficient [10, p. 248]) is an evidence of a good mutual consistency of the RMs within the set. The proximity of the angular coefficient of the graph (equal to 0.96) to 1 indicates a rational method of analytical signal generation. A comparison with the calibration dependence obtained under similar conditions for the RM set

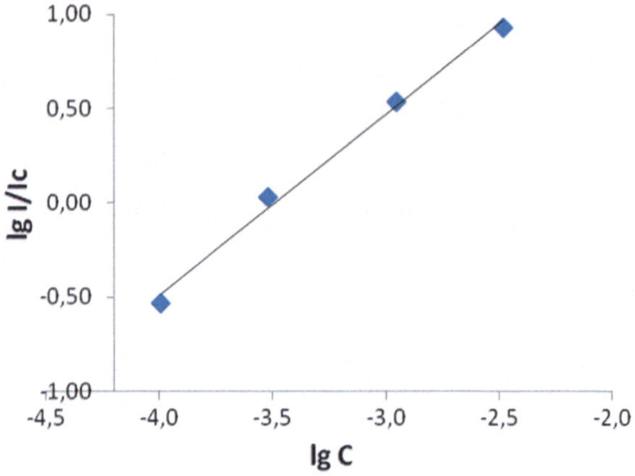

Fig. 2 Calibration graph, reproduced for the SOBor set

produced by the MSZ laboratory described by the equation y = (0.91 ± 0.14) x + (3.07 ± 0.46), indicates the absence of significant differences between the values of the coefficients in both graphs.

The correctness of the calculated values of the boron content in the developed RMs has been confirmed by the results of SOBor material analysis in the calibration of a spectrometer using a set of in-house reference materials. A comparison of the obtained measurement results with the certified values of the mass fraction of boron was performed by regression analysis [15] when plotting the dependence in logarithmic coordinates due to the wide range of variation of the element concentration in the RM samples, covering 1.5 orders of magnitude. The type of the initial connection between the experimental (C) and certified ($\hat{C}$) values of the element concentration is taken to be $C = k\hat{C}^a$. The transition to logarithmic coordinates leads to the transformation of the model into a linear equation

$$\lg C = b + a \lg \hat{C}, \tag{1}$$

where a is the slope, b is a free member and $b = lgk$. The test is based on estimating the significance of differences between the regression coefficients and the ideal values: the slope $a = 1.11$ from 1 and the free term $b = 0.41$ from zero using Student's t-distribution. The calculations show that the statistics generated for both coefficients equal to 2.47 and 2.72 do not exceed the quantile of the distribution $t_{(0.975;2)} = 4.30$ [10, p. 178]. Therefore, the coefficients a and b can be taken equal to 1 and 0, respectively, which is a proof of the correctness of the established values of the concentration of the component being certified. This conclusion confirms a negligible difference of the determination coefficient from 1, since $R^2 = 0.997 > r^2 = 0.903$, where $r_{(0.975;2)}$ is the correlation coefficient [10, p. 248].

The metrological traceability of the certified RM values to the mass units was established using calibrated balances [9], as well as certified gravimetric [8] and atomic emission [2] measurement methods.

Table 4 Certified values of the mass fraction of boron (W, %), limits of the absolute ($\pm\Delta$, %) and relative ($\pm\delta$, %) errors

RM	W (%)	$\pm\Delta$ (%)	$\pm\delta$ (%)
SOBor-1	0.00323	0.00004	0.2
SOBor-2	0.001087	0.000019	1.7
SOBor-3	0.000299	0.000007	2.4
SOBor-4	0.000102	0.000005	4.6

The mass fraction of boron calculated relative to the sum of the masses of aluminium, silicon and nickel in silumin is found to range from 1×10^{-4} to 3.3×10^{-3}% with a relative error (±δ) not exceeding 5% (Table 4). The expiry date of the developed RMs estimated by the method of accelerated aging under the conditions of increased thermal and vibration loads is set to be 60 years.

3 Conclusions

According to the obtained metrological characteristics of the developed RMs, these materials meet all the requirements specified both in the technical specification to the development of an RM set and RF regulatory documents for the manufacture and characterisation of RMs of the substances composition. The developed CRM set is included in the State Register of Approved Reference Material Types under the number GSO 11178-2018/11181-2018.

Author Contribution Statement All authors contributed equally to this work.

References

1. Lisienko DG, Dombrovskya MA (2005) Certified reference materials of graphite collector of microimpurities: synthesis, certification, application. Analitika i Control 9(3):285–294 (In Russ.)
2. Lisienko DG, Dombrovskya MA, Kubrina ED, Varkentin NY (2016) Material synthesis and evaluation of metrological characteristics of potassium fluozirconate certified reference material. Ref Mater 3:47–60. https://doi.org/10.20915/2077-1177-2016-0-3-47-60 (In Russ.)
3. MI 1992-98 GSI National measurement standards—Universal analyzers of composition of substances and materials—General requirements for in-service verification methods. Yekaterinburg (In Russ.)
4. OI 001.339-2006 Bor (2016) Atomic emission method measurements in silumin, intermetallic, electrocorundum. Electrostal, Public Joint-Stock Company Mashinostroitelny Zavod (In Russ.)
5. Budanova LM, Volodarskaya RS, Kanaev ON (1966) Analysis of aluminum and magnesium alloys. Metallurgy, Moscow (In Russ.)
6. Tikhonov VN (1971) Analytical chemistry of elements: aluminum. Nauka, Moscow
7. GSO 4519-89/4523-89 RM of the composition of graphite (set SOG-21). https://fgis.gost.ru/fundmetrology/registry/19/items/393181

8. Instruction No. 002-SO-2008 10-water recrystallized sodium tetraborate. Methods of measuring the mass fraction of sodium tetraborate anhydrous by the gravimetric method. Yekaterinburg, AIC-RAL UrFU (2016) (In Russ.)
9. GOST R 53228-2008 (2010) Non-automatic weighing instruments. Part 1. Metrological and technical requirements. Tests. Standartinform, Moscow (In Russ.)
10. Bol'shev LN, Smirnov NV (1983) Tables of mathematical statistics. Nauka, Moscow (In Russ.)
11. GOST 8.531-2002 (2003) State system for ensuring the uniformity of measurements. Reference materials of composition of solid and disperse materials. Ways of homogeneity assessment. Standartinform, Moscow (In Russ.)
12. MI No. 020-SO-2014 (2014) Methods for measuring the ion counting rate (ionic currents) in aqueous solutions by mass spectrometry method. AIC-RAL, UrFU, Yekaterinburg (In Russ.)
13. R 50.2.031-2003 GSI (2004) Reference materials of composition and properties of substance and materials. Procedure to evaluate the stability. State Standard Moscow 2004. (In Russ.)
14. Lisienko DG, Ulyanova NV, Dombrovskaya MA (1995) Sustainability assessment of state reference material materials of zirconium oxide composition. VINITI RAN, Moscow (In Russ.)
15. Nalimov VV (1960) Application of mathematical statistics in the analysis of substance. Moscow (In Russ.)

Preparation and Certification of a Reference Material for the Physical Properties of Clay Soils (Loam)

Ilona N. Matveeva and Vladimir V. Tolmachev

Abstract This paper presents information on the development of a reference material (RM) for the physical properties of clay soils. The list of certified characteristics included the liquid limit determined by the fall cone test, the plastic limit and the particle density determined by the pycnometric method. Establishment of the certified values of clay soil physical properties was performed by the method of interlaboratory comparisons. Absolute uncertainty interval for the certified values of liquid limit, plastic limit and soil particle density are found to be 1.9%, 1.5% and 0.03 g/cm^3, respectively. The shelf life of the certified reference material is determined to be 5 years. The developed certified reference material is included into the State Register of the Approved Types of CRMs as CRM (GSO) 11038-2018. The CRM can be used for controlling the accuracy of certified characteristics measurements, conducting interlaboratory comparisons and monitoring the performance of individual laboratories.

Keywords Clay soil (loam) · Soil physical properties · Liquid limit · Plastic limit · Particle density · Certified reference material

Abbreviations

CRM Certified reference material
ILCs Interlaboratory comparisons

I. N. Matveeva (✉) · V. V. Tolmachev
Ural Scientific Research Institute for Metrology (UNIIM), Ekaterinburg, Russia
e-mail: sertif@uniim.ru

S. Medvedevskikh et al. (eds.), *Reference Materials in Measurement and Technology*,
https://doi.org/10.1007/978-3-030-32534-3_8

1 Introduction

Engineering surveys are known to be an important part of any construction process and must precede the stage of planning and design. Proper engineering of supporting structures allows the defects and disruptions of buildings associated with the subsidence of the foundation to be avoided, in accordance with the RF Federal Law No. 384-FZ dated 30.12.2009 [1]. In addition, structural engineering is supposed to ensure the safety of roads according to the Technical Regulations of the Customs Union, TR CU 014/2011 [2].

When conducting engineering surveys, an accurate determination of soil characteristics is essential for informing design and construction decisions. According to GOST 25100 [3], soils differ in terms of numerous characteristics. Physical properties of soils characterise the physical state of a particular soil and its ability to change the state under the influence of various physical and chemical factors (hydrogeological and thermal conditions, load deformation, etc.).

A variety of special equipment enables survey organisations to carry out soil research directly at construction sites. However, in situ soil testing allows exclusively the determination of engineering and geological conditions of a land object, such as soil composition, its thickness, soil frost line, presence of groundwater and aquifers, etc. In order to determine the physical properties of soils, laboratory studies are required.

The territory of the Russian Federation predominantly features sand and clay soil types. Clay soils (loam) consist mainly of dust and clay particles and appear to be among the most problematic ones for construction. These soils are characterised by plasticity, i.e. the capability of modifying their state when moisturised or frozen (from solid to semi-solid, then to plastic and liquid). Therefore, a detailed study of clay soils in terms of moisture attributes allows accurate prediction of changes in their state under various geological conditions and operation.

2 Literature Review

The types and content of laboratory methods used to determine soil characteristics are defined in Appendix M of the Design and Construction Specification SP 11-105 [4] and GOST 30416 [5]. At the same time, different methods can be applied to determine a particular soil characteristic, depending on the soil type and target application. For example, the determination of soil particle density can be carried out using the water pycnometric method for all soil types, the neutral liquid pycnometric method for saline and swelling soils, and two pycnometers for saline soils.

Laboratory methods for determining the physical characteristics of soils are established in GOST 5180 [6]. The main physical characteristics to be determined are moisture, density, composition, permeability, deformability and strength.

When conducting laboratory experiments on the physical properties of soils, testing laboratories should not only be able to establish the reliable values of the investigated properties, but also to monitor the accuracy of the conducted research. The results obtained by laboratory research, as well as their accuracy and reliability, should agree with the parameters established in the project of engineering geological surveys.

The current practice of testing laboratories implies the accuracy control by increasing the number of parallel experiments, which undoubtedly raises the cost and complexity of laboratory studies. However, the quality of research into the physical properties of soils can be ensured by applying certified reference materials (CRMs) for every soil type.

3 Materials and Methods

An RM for the composition of clay soils was developed from loam collected at the Reshetnikovskoye oilfield, Udmurt Republic, Mozhginsky region (Russia).

The technological preparation of the starting material included the following operations, necessary to ensure the homogeneity of the obtained RM:

1. Manual selection of rectangular-shaped monoliths from the soil mass of pits in the selected area. The sampling depth was (2.1–2.3) m;
2. Preliminary grinding of the soil samples followed by air drying in a ventilated room;
3. Final grinding followed by sieving through a calibrated sieve with a 1 mm mesh. All inclusions (gravel, nodules and organic residues) were removed;
4. Final drying followed by dividing up to the required weight of 250 g using the quartering method. Subsequent packaging of the finished samples into rectangular plastic bags using a vacuum machine.

The granulometric composition of the obtained powder was determined by the hydrometer test using a hydrometer and a set of sieves with a mesh size of 10, 5, 2, 1.0, 0.5, 0.25 and 0.1 mm. The obtained values of the content of various-size particles in the CRM fractions are presented in Table 1.

Eventually, the RM represented a powder with a particle size of less than 1 mm, prepared from loam soil, ground and dried to an air-dried basis, containing no organic inclusions and packaged per 250 g into hermetic plastic bags.

3.1 Homogeneity Study

The obtained CRM was examined for homogeneity.

Table 1 Granulometric composition of the RM under development

Particle size (mm)	Particle content (% wt)
1.00–0.50	<0.5
0.50–0.25	<0.5
0.25–0.10	1.5–3.0
0.10–0.05	4.0–9.0
0.05–0.01	46.0–54.0
0.002	18.0–24.0
<0.002	16.0–24.0

The homogeneity study was carried out in accordance with GOST 8.531 [7] using an algorithm for dispersed materials. The measurements were conducted using 130 samples, with 2 parallel determinations being performed for each sample.

Technical requirements for the development of the CRM set the values of the uncertainty induced by inhomogeneity for the liquid and plastic limits to be lower than 1.0%, and the density of soil particles to be lower than 0.015 g/cm^3. The interval of the uncertainty induced by inhomogeneity was established on the basis of data on the repeatability limit of the results of parallel determinations given in GOST 5180 [6].

For the RM under study, the values of the uncertainty induced by inhomogeneity for liquid and plastic limits and soil particle density were established to be equal to 0.9%, 0.6% and 0.01 g/cm^3, respectively. The obtained values are shown to be in good agreement with the formulated technical requirements for the CRM development; therefore, the RM is recognised to be homogeneous.

3.2 *Stability Study*

The stability of the certified values of the RM during the expected shelf life was confirmed by experimental studies conducted over a five-year period on an experimental batch using appropriate measurement methods in accordance with GOST 5180-2015 [6].

The study of the CRM stability was carried out in compliance with a method for evaluating stability characteristic specified in R 50.2.031 [8]. According to the results of the stability study, the developed CRM is assigned with the shelf life of 5 years for all the certified characteristics.

3.3 *Determination of the RM Certified Values and Respective Uncertainties*

The determination of the RM certified values and the estimation of respective uncertainties was carried out according to GOST 8.532 [9] using the method of interlaboratory comparisons (ILCs).

Fifteen independent laboratories accredited for technical competence in the National accreditation system and having experience in soil research participated in the undertaken ILCs.

The determination of the certified values was carried out by standardised methods provided in GOST 5180-2015 [6]. The fall cone test, plastic limit test and water pycnometric method were applied by all the participants to determine the liquid limit, the plastic limit and the density of soil particles, respectively.

The uncertainties of the certified values were established taking into account the contribution from the inhomogeneity of the CRM.

The developed CRM is included into the State Register of Approved Types of CRMs as the RCM (GSO) 11038-2018.

4 Results

The values of the metrological characteristics of the CRM (GSO) 11038-2018 are presented in Table 2.

The traceability of the certified values obtained during the ILC to the respective SI units is ensured through the use of verified measurement instruments and standardised measurement techniques by testing laboratories with experience in the study of substances, similar or equivalent by state and structure, to the CRM under research, and accredited in the Russian national accreditation system.

The CRM for the physical properties of clay soils (loam) is intended for monitoring the accuracy of experiments used to measure soil liquid limits by the fall cone test, soil plastic limits and soil particle densities using the pycnometric method, all performed according to GOST 5180-2015 [6].

Table 2 Metrological characteristics of the CRM (GSO) 11038-2018

Certified characteristic	Unit	Certified value	Absolute uncertainty interval of the certified value at $P = 0.95$, $\pm\Delta$ (%)
Liquid limit by the fall cone test	%	36.7	1.9
Plastic limit	%	21.2	1.5
Density of soil particles by the pycnometric method	g/cm3	2.71	0.03

The developed CRM can be applied for both conducting interlaboratory comparisons and monitoring the proficiency of individual laboratories.

5 Discussion and Conclusions

The CRM for the physical properties of clay soils (loam), which has no analogues in the Russian Federation, has been developed and certified. This CRM can be used to control the accuracy of liquid limit measurements via the fall cone test, plastic limit and the density of soil particles using the pycnometric method.

The development of CRMs for other physical properties and other types of soil appears to be a priority task for future research.

Author Contribution Statement I. N. Matveeva: research concept, collection of experimental data, analysis of experimental information, literature review, manuscript preparation,

V. V. Tolmachev: concept of research, critical analysis and revision of the article.

References

1. RF Federal Law N 384-FZ of 12.30.2009 (2009) Technical regulation on safety of buildings and structures. Adopted by the State Duma on 23 Dec 2009 (In Russ)
2. TR CU 014/2011 Customs Union Technical Regulations on safety of automobile roads. Adopted by the Decision of the Commission of the Customs Union of 18.10.2011 No 827 (In Russ)
3. GOST 25100–2011 Soils (2013) Classification. Standartinform, Moscow, 42 p (In Russ)
4. SP 11-105-97 (1997) Engineering and geological surveys for construction. Part I. General rules for the production of works. PNIIIS Gosstroy of Russia (In Russ)
5. GOST 30416–2012 (2013) Soils. Laboratory testing. General requirements. Standartinform, Moscow, 16 p (In Russ)
6. GOST 5180–2015 Soils (2016) Laboratory methods for determination of physical characteristics. Standartinform, Moscow, 24 p (In Russ)
7. GOST 8.531–2002 (2002) State system for ensuring the uniformity of measurements. Reference materials of composition of solid and disperse materials. Ways of homogeneity assessment. Izdatelstvo Standartov, Moscow, 15 p (In Russ)
8. R 50.2.031–2003 (2004) State system for ensuring the uniformity of measurements. Certified reference materials of composition of substances and materials. Ways of stability characteristics assessment. Izdatelstvo Standartov, Moscow, 12 p (In Russ)
9. GOST 8.532–2002 (2008) State system for ensuring the uniformity of measurements. Certified reference materials of composition of substances and materials. Interlaboratory metrological certification. Content and order of works. Standartinform, Moscow, 12 p (In Russ)

Development of Reference Materials for Thermodynamic Properties: Metrological Support of Measurements in the Field of Thermal Analysis and Calorimetry in Russia

Andrey M. Nepomiluev, Vyacheslav V. Kazantsev and Artyom P. Shipitsyn

Abstract This paper is aimed at analysing the current state and prospects of metrological support and standardisation in the field of thermal analysis in Russia. Main characteristics of reference materials used for testing, graduating, calibrating and verifying thermal analysis instruments are described.

Keywords Reference materials · Thermal analysis · Specific enthalpy · Specific heat · Phase transition heat

1 Introduction

Thermal analysis (TA) is one of the most dynamically developing methods for studying the structure and properties of various substances. Due to the rapid expansion of TA applications in science and industry, TA methods are widely used in all areas of chemistry, particularly in polymer chemistry. However, requirements for the performance of thermo-analytical equipment are continuously increasing in terms of both enhancing its sensitivity and accuracy while reducing measurement time. The required metrological characteristics cannot be achieved without suitable certified reference materials (RMs).

2 Literature Review

At present, TA refers to a whole group of methods aimed at measuring the physical properties of substances as a function of temperature or time. During such testing, the analysed substances are subjected to a controlled temperature program [1]. The temperature program can include heating, cooling at a constant rate, holding at a constant temperature (isotherm), a combination of these modes or a temperature

A. M. Nepomiluev · V. V. Kazantsev (✉) · A. P. Shipitsyn
Ural Scientific Research Institute for Metrology (UNIIM), Ekaterinburg, Russian Federation
e-mail: kazantsev@uniim.ru

S. Medvedevskikh et al. (eds.), *Reference Materials in Measurement and Technology*,
https://doi.org/10.1007/978-3-030-32534-3_9

modulation mode with a specified amplitude and frequency. The classification of main TA methods [1] is given in Table 1.

Such TA methods as differential thermal analysis (DTA) and differential scanning calorimetry (DSC) are currently among the most advanced and highly sensitive methods for the phase analysis of heterogeneous systems. These methods allow the determination of not only the temperature and enthalpy of phase transformations, the heat capacity and its dependence on thermodynamic parameters, but also the kinetic characteristics of physicochemical processes under the conditions of linear temperature variation.

Such an extensive list of experimental methods combined under the umbrella term "thermal analysis" has resulted in the introduction of a large range of measuring instruments (MIs) into the Federal Information Fund for Ensuring the Uniformity of Measurements by leading manufacturers of thermo-analytical equipment, such as Netzsch-Geratebau GmbH (Germany), TA Instruments and PerkinElmer, Inc. (USA), Setaram Instrumentation (France) and Mettler-Toledo AG (Switzerland). As of 2018, the number of MIs in the field of TA amounted to 130 models (Table 2). The undoubted leader among MI producers is Netzsch-Geratebau GmbH (Germany) accounting for 34% (44 models) of MIs entered in the MI State Register and occupying about 70% of the market for thermo-analytical equipment in Russia.

Out of 129 type-approved MI models, there are 40 types of differential scanning calorimeters alone, covering a fairly wide range of temperatures from minus 180 to 1650 °C. The market has recently seen the appearance of a number of high-temperature devices with the maximum operating temperatures of 2000 °C (Netzsch DSC 404 F1/F3 Pegasus, Netzsch STA 449 F1 Jupiter, Germany), 2400 °C (Netzsch STA 449 F3 Jupiter, Germany; Setaram Instrumentation SETSYS Evolution TGA-DTA/DSC, France) and even 2800 °C (Netzsch DIL 402 Expedis HT, Germany).

Reference materials (RMs) for thermophysical properties are a conventional means of ensuring the uniformity of measurements in various types of TA. Since some of the main parameters measured during the TA of various substances are temperature, enthalpy (heat) of phase transformations and specific heat, MI calibration and verification by temperature and enthalpy are carried out using RMs for temperature and heat of fusion. At the same time, it is necessary to have different types of RMs based on materials with different temperature and enthalpy of phase transformations. Leading producers of MIs for TA provide a wide range of substances for calibrating and graduating measuring cells made of various materials. For example, the list of calibration samples of the world leader in the production of thermo-analytical equipment Netzsch Geratebau GmbH (Germany), consists of 24 items and covers almost the entire range of temperature and specific enthalpy values. This list includes both metals and non-metals, with 12 items of each type being presented (Table 3). However, not all substances listed in Table 3 are certified as RMs—GSO (or standard reference materials—SRMs).

The Ural Scientific Research Institute for Metrology (UNIIM), a competent producer of RMs for thermophysical properties, has developed 9 RM types for thermophysical properties, including those based on corundum, potassium chloride, stainless steel, gallium, indium, tin, zinc, antimony and molybdenum (Table 4). Metrologi-

Table 1 Classification of main TA methods

Method name	Method definition	Purpose of the method
Differential thermal analysis (DTA)	The method is used to measure the temperature dependence of the temperature difference between the substance and the reference subjected to a temperature program	The method studies the thermal effects of physical and chemical processes: temperature and enthalpy of phase transitions and chemical reactions; specific heat; thermostability. It also provides quantitative and qualitative analysis of the phase composition of a substance
Differential scanning calorimetry (DSC)	The method is used to measure the temperature dependence of heat-flow difference between the substance and the reference material subjected to a temperature program	
Thermogravimetric analysis (TGA)	The substance mass is measured as a function of temperature or time, while the substance is exposed to a controlled temperature in a given atmosphere	• Measurement of mass changes resulting from: the sorption/desorption of a substance from the surface or from the sample volume; the evolution of gases in the process of pyrolysis; redox reactions occurring when the sample starts interacting with the atmosphere • Determination of humidity, thermal stability and the chemical composition of a substance
Thermomechanical analysis (TMA)	The deformation of a substance under a non-oscillating load is measured as a function of temperature or time, while the substance is exposed to a temperature program in a given atmosphere	• Study of second-order phase transitions • Study of the viscoelastic properties of substances and materials; the characteristics of stiffness and damping; resizing under static and dynamic (oscillating and non-oscillating) load • Study of glass transitions in organic and inorganic systems
Dynamic mechanical analysis (DMA)	The elastic and loss moduli of a substance under an oscillating load are measured as a function of temperature, time or oscillation frequency, while the substance is exposed to a temperature program in a given atmosphere	

(continued)

Table 1 (continued)

Method name	Method definition	Purpose of the method
Dilatometry (DIL)	The size of a sample being under a negligible load is measured as a function of temperature, while the sample is subjected to a temperature program in a given atmosphere	• Measurement of dimensional changes of solid, liquid, pasty substances and powders at a programmed temperature change with negligible load on the sample • Determination of (linear and volumetric) thermal expansion coefficients
Dielectric analysis (DEA)	The dielectric constant (dielectric permittivity or capacitance) and the dielectric loss (conductivity) of a substance in an oscillating electric field are measured as a function of temperature or time, while the sample is subjected to a temperature program in a given atmosphere	• Determination of dielectric constant, loss factor, conductivity, resistance (ion viscosity); coefficient of curing (degree of crosslinking) of polymers

Table 2 Main manufacturers of thermo-analytical equipment

Manufacturer	Number of type-approved MIs as of 2018
Netzsch-Geratebau GmbH (Germany)	44
TA Instruments (USA)	29
PerkinElmer, Inc (USA)	16
Setaram instrumentation (France)	19
Mettler-Toledo AG (Switzerland)	21
Total	129

cal services of more than 300 enterprises, research organisations and educational institutions apply these RMs in their everyday practice. The development and certification of the RMs was performed in accordance with the State verification scheme for instruments measuring specific enthalpy and specific heat of solids in the temperature range 700–1800 K [2], using the GET 67-2013 State Primary Special Measurement Standard for units of specific enthalpy and specific heat of solids in the temperature range 700–1800 K [3].

Table 3 List of RMs supplied by Netzsch for calibrating (graduating) TA instruments

Substance	Melting point (of phase transition)		Enthalpy (heat) of phase transition (J/g)	Note
	°C	K		
C_6H_{12}	−87.0	186.0	−79.4	
$C_{10}H_{16}$	−64.7	208.3	−22.3	
Hg	−38.8	234.2	−11.4	
H_2O	0.0	273.0	−333.4	
Ga	29.8	302.8	−80.0	
$C_{12}H_{10}$	69.2	342.2	−120.5	
C_6H_5COOH	122.4	395.4	−147.4	
KNO_3	128.7	401.7	−50.0	Only for metal crucibles
In	156.6	429.6	−28.6	
$RbNO_3$	164.2	437.2	−26.6	Only for metal crucibles
Sn	231.9	504.9	−60.5	Only for ceramic crucibles
Bi	271.4	544.4	−53.1	Only for ceramic crucibles
$KClO_4$	300.8	573.8	−104.9	Only for metal crucibles
Pb	327.5	600.5	−23.0	Only for ceramic crucibles
Zn	419.5	692.5	−107.5	Only for ceramic crucibles
Ag_2SO_4	426.4	699.4	−51.0	Only for metal crucibles
CsCl	476.0	749.0	−17.2	Only for metal crucibles
Al	660.3	933.3	−397.0	Only for ceramic crucibles
K_2CrO_4	668.0	941.0	−37.0	Only for metal crucibles
$BaCO_3$	808.0	1081.0	−94.9	Only for metal crucibles
Ag	961.8	1234.8	−104.6	Only for ceramic crucibles
Au	1064.2	1337.2	−63.7	Only for ceramic crucibles
Ni	1455.0	1728.0	−290.4	Only for ceramic crucibles
Pd	1554.8	1827.8	−157.3	Only for ceramic crucibles

The data analysis given in Tables 3 and 4 shows that the list of materials supplied by the manufacturers of thermo-analytical equipment for calibrating instruments is much wider than that of GSO. A similar situation is characteristic not only of Russia, but of other countries as well. For example, the list of RMs for thermophysical properties offered by NIST (USA) also includes 9 items (Table 5), while the list of materials recommended [4] for the melting-point calibration of thermo-analytical equipment comprises 20 items and includes such materials as Pt (T_m = 1772 °C) and Rh (T_m = 1963 °C).

Further development of thermal methods for analysing materials follows several directions. One of the main trends in the development of TA methods has so far been enhancing the accuracy of techniques aimed at measuring the heat capacity substances, as well as temperatures and enthalpies of physico-chemical transformations.

Table 4 List of GSO for thermophysical properties

State register number, RM material	Certified characteristic	Certified characteristic range	Error (±)
GSO 149-86P, corundum	Specific enthalpy	−96.2 to 2477.3 kJ/kg	(0.1–0.3) %
	Specific heat	0.09–1.4 kJ/(kg K)	(0.1–1.5) %
GSO 886-76, stainless steel	Specific enthalpy	51.6–629.4 kJ/kg	1%
	Specific heat	0.49–0.65 kJ/(kg K)	1.5%
GSO 10898-2017, molybdenum	Specific enthalpy	107.6–456.4 kJ/kg	(0.1–0.3) %
	Specific heat	0.28–0.36 kJ/(kg K)	(0.3–0.8) %
GSO 1363-78, potassium chloride	Specific enthalpy	121.8–541.6 kJ/kg	0.3%
	Specific heat	0.73–0.84 kJ/(kg K)	1%
	Specific heat of fusion	357.29 kJ/kg	0.5%
	Melting point	1044.75 K	0.6 K
GSO 2312-82, gallium	Melting point	303.04 K	0.06 K
GSO 2313-82, indium	Melting point	429.85 K	0.06 K
	Specific heat of fusion	28.58 kJ/kg	0.12 kJ/kg
GSO 2314-82, tin	Melting point	505.20 K	0.12 K
	Specific heat of fusion	59.92 kJ/kg	0.25 kJ/kg
GSO 2315-82, zinc	Melting point	692.71 K	0.26 K
GSO 2316-82, antimony	Crystallisation temperature	903.76 K	0.03 K

Table 5 List of reference materials (SRMs) for thermophysical properties (NIST)

Number	Name
SRM 720	Synthetic sapphire
SRM 2232	Indium
SRM 2220	Tin
SRM 2234	Gallium
SRM 781D2	Molybdenum
SRM 2235	Bismuth
SRM 2225	Mercury
SRM 705a	Polystyrene

Another equally important task is to increase the rate, sensitivity and, consequently, the temperature resolution of TA equipment. Thus, a gradual expansion of the temperature measurement range occurs. All the abovementioned leads to the conclusion that the range of GSO (SRMs) used for testing, graduation, calibration and verification of thermo-analytical equipment should be expanded. Along with that, RMs must meet the following basic requirements:

- high degree of purity (not less than 99.999% of the main substance);
- well-established phase transition characteristics;
- resistance to air and radiation;
- chemical stability across the temperature range and under given operation conditions;
- low pressure of saturated vapour at the phase transition temperature;
- inertness with respect to the crucible material across the temperature range and under given operation conditions;
- thermophysical (heat capacity, thermal conductivity) and physical (mass, dimensions) characteristics related to the materials under study.

3 Conclusion

The active development of TA methods and the sustainable growth of thermoanalytical equipment require a significant expansion of the thermal-property RM range based on the recommendations outlined in [5–7].

References

1. ASTM E 473-16 Standard terminology relating to thermal analysis and rheology
2. GOST R 8.872-214 State verification schedule for means of measuring specific enthalpy and specific heat of solids in the range of temperature from 700 to 1800 K (In Russ)
3. Kazantsev VV, Cherepanov VI, Sennikova VN, Averkiev MV (2015) National primary special standard GET 67-2013 for the units of specific enthalpy and specific heat of solids at temperatures in the 700–1800 K range. Meas Tech 58(2):121–126 (In Russ)
4. ASTM E 967-08 Standard test method for temperature calibration of differential scanning calorimeters and differential thermal analyzers
5. ISO 11357-1: 2016 Plastics—differential scanning calorimetry (DSC)—Part 1: general principles
6. Sabbah R, An Xu-Wu, Chickos JS, Planas Leitão ML, Roux MV, Torres LA (1999) Reference materials for calorimetry and differential thermal analysis. Thermochim Acta 331:93–204
7. Gatta GD, Richardson MJ, Sarge SM, Stolen S (2006) Standard and calibration and guidelines in microcalorimetry. Part 2. Calibration standards for differential scanning calorimetry (IUPAC Technical Report). Pure Appl Chem 78(7):1455–1476

Development of Alumina-Based Porosity Reference Materials for the Mercury Porosimetry Method

Egor P. Sobina

Abstract This paper presents the results of a study on the development of alumina-based porosity reference materials (RMs), which are intended for controlling the accuracy of the measurement results obtained using the mercury porosimetry method. These mercury porosimetry RMs are being developed in the Russian Federation for the first time and are primarily intended for verification/calibration and testing for type approval of mercury porosimeters. The study of metrological characteristics was conducted using a reference system, which is to be included in the GET 210-2014 State Primary Measurement Standard for units of specific gas adsorption, specific surface area, specific volume and pore size of solid substances and materials. This reference system was calibrated prior to conducting the study with the use of external reference pressure sensors of the 2nd order at 0.6 and 400 MPa, as well as a laboratory balance (special accuracy class I) having a maximum weighing limit of 220 g. In the course of conducted research work, procedures for reproducing and estimating the uncertainty of porosity characteristics using the mercury porosimetry method were developed. When estimating the RM metrological characteristics, the standard uncertainty, associated with the characterisation and heterogeneity of the studied materials (RM candidates), was evaluated. The standard uncertainty due to instability was disregarded due to the high stability of the alumina matrix. According to the results of the conducted studies, four types of alumina-based porosity RM were developed. Their pore sizes are in the region of ~10 nm; ~2,000 nm; ~9,000 nm; ~60,000 nm. The RMs with pore sizes of ~10 nm; ~2,000 nm were intended for use in controlling the accuracy of mercury porosimeters operating at high pressures, while the RMs with pore sizes of ~9,000 nm; ~60,000 nm were for controlling the accuracy of mercury porosimeters operating at low pressures. The developed RMs are different from their foreign counterparts, which are produced by BAM (Germany) and NIST (USA) metrological institutes. The standard uncertainty due to characterisation of the RMs in question was estimated, not by the method of interlaboratory experiment, but by a full analysis of the constraint equations taking into account all sources of uncertainty included in the equations on the basis of the Guide to the Expression of Uncertainty in Measurement. The conducted studies allowed the RM certified values

E. P. Sobina (✉)
Ural Scientific Research Institute for Metrology (UNIIM), Ekaterinburg, Russia
e-mail: 251@uniim.ru

S. Medvedevskikh et al. (eds.), *Reference Materials in Measurement and Technology*,
https://doi.org/10.1007/978-3-030-32534-3_10

to be traceable at this stage to the GET 3-2008 State Primary Measurement Standard for the mass unit (kilogram), the GET 101-2011 State Primary Measurement Standard for the pressure unit in the field of absolute pressure in the range 1×10^{-1}–7×10^{5} Pa, the GET 43-2013 State Primary Measurement Standard for the pressure unit in the range 10–1600 MPa and the effective area of the piston-cylinder unit of dead weight testers in the range 0.05–1 cm^2. In addition, a reference material having a large pore size of ~60,000 nm, which currently has no analogues elsewhere in the world, was characterised.

Keywords Mercury porosimetry · Porosimeter · Reference material · Metrological traceability · Pore size · Specific pore volume · State primary measurement standard

1 Introduction and Literature Review

The creation of new porous materials, whose quality must be controlled, necessitates the introduction of a wide range of instruments for measuring porosity characteristics. In particular, the pore size significantly affects the performance of ceramic filter elements, which are intended for separating, concentrating and purifying various liquid mixtures from solid particles. One of the most accurate methods for measuring pore size within the range from 100 nm to 0.9 mm is the mercury porosimetry method [1–3]. This method is characterised by its great versatility, since it allows porous structure information to be obtained across a wide range of pore sizes, and its calculation equations are distinguished by their simplicity [1–3]. It is also possible to use this method for measuring the specific surface area of dispersed bodies under the following conditions: (1) when the powder has a relatively low surface energy and mercury does not wet the surface of its particles; (2) when these bodies have no blind pores and the pressure in the porosimeter ensures the penetration of mercury into the smallest micropores of the material [3]. The objects of application of mercury porosimetry are rocks and ores, pigments, carbon and graphite, catalysts and minerals, glass, coatings, adsorbents, fertilizers, coal, filters, fillers, fabrics and fibres, gaskets, fibrous plates, refractories, paper, nuclear fuel, metal slag and composite materials. Mercury porosimeters are used to determine the total pore volume and pore size distribution, specific surface area, average pore diameter, bulk density and true density. The mercury porosimetry method is based on the fact that mercury, which does not wet a solid, penetrates into its pores only when external pressure is applied. However, there are currently no measuring instruments or corresponding reference materials (RMs) of approved type being produced in Russia based on the mercury porosimetry method. Therefore, in order to control the accuracy of mercury porosimetry results, Russian enterprises have to use foreign RMs [4]: either (1) materials for which the uncertainty of certified values has not been estimated and the measurement traceability has not been established, or (2) certified reference materials (CRMs), which are costly. Abroad, RMs are produced by BAM

(Germany) [5] and NIST (USA) metrological institutes [6]. The metrological institute BAM (Germany) produces nine RMs of various porous materials: macroporous membranes, macroporous ceramic materials based on aluminium oxide and porous glasses in the following ranges: pore size ranging from 15 to 6000 nm with a relative expanded uncertainty from 1.3 to 11.7%, specific pore volume ranging from 110.9 to 919.7 mm^3/g with a relative expanded uncertainty from 1.8 to 3.9% [5]. NIST (USA) produces one extruded compound RM based on silica and aluminium oxide having the following metrological characteristics and certified values: (1) pore size of 8.847 nm, with a relative expanded uncertainty of 4.1%; (2) specific pore volume of 0.547 mm^3/g, with a relative expanded uncertainty of 3.3% [6].

Therefore, this work is devoted to the development of alumina-based porosity RMs for the metrological assurance of the mercury porosimetry method.

2 Materials and Methods

The following source materials for creating RMs were chosen:

- 3 different types of ceramic filter elements produced by the scientific production association Ceramic filter LLC (Moscow) in conformity with TU 3614-001-18985634-2006, comprised of alumina-based macroporous membranes;
- granular aluminium oxide produced by Ishimbay Specialized Chemical Plant of Catalysts LLC (Ishimbay) in conformity with TU 2163-015-94262278-2009.

The pore size of these materials varies over a wide range from 10 nm to 60,000 nm. The metrological characteristics of candidate RMs are studied by means of a reference system implementing the mercury porosimetry method. This system is to be included in the GET 210-2014 State Primary Measurement Standard for units of specific gas adsorption, specific surface area, specific volume and pore size of solid substances and materials [7] at the end of 2019. It consists of the following equipment: (1) a Thermo Fisher Scientific mercury porosimeter (consisting of two stations: Pascal 140 EVO low pressure station and Pascal 440 EVO high pressure station) providing a range of pore size measurements from 3.6 to 9×10^5 nm; (2) a mercury purification apparatus; (3) an apparatus for calibrating the volume of mercury that intrudes into the material; (4) a set of dilatometers for measuring compact and granular materials; (5) a personal computer with specialized SOLID EVO software; (6) an A&D GX-600 laboratory balance (special accuracy class I); (7) a Pycnomatic ATC gas pycnometer (Thermo Fisher Scientific) capable of maintaining a set temperature within a range from 18 to 35 °C and measuring the density of solids and materials within a range from 2 to 23 g/cm^3; (8) a LR-Cal LPC 200 pressure calibrator with LPC-2-6 and LPC-2-4000 absolute pressure sensors (DRUCK & TEMPERATUR Leitenberger GmbH) having measurement limits of 600 kPa and 400 MPa, with permissible absolute pressure fiducial error limits of ±0.025% and ±0.1%, respectively; (9) a EKON UKR-1MC universal mercury measuring apparatus of

Research-and-production environmental firm for determining mercury vapour in ambient air.

The mercury porosimetry method is based on the fact that mercury, which does not wet a solid, penetrates into its pores only when external pressure is applied. The way a liquid behaves in porous materials depends on its interaction with the solid substance, i.e. on the action of surface forces. This interaction is characterised by the contact angle (wetting contact angle) θ. If $\theta < 90°$, the liquid's behaviour with respect to this solid is called wetting; if $\theta > 90°$, then it is called non-wetting. A wetting meniscus has a concave shape in the capillaries. The liquid below the meniscus is under negative pressure, whose value is expressed by its rise in the capillary. Conversely, non-wetting meniscus has a convex shape and the liquid underneath it is under positive pressure, which is measured by a lowering of the liquid in the capillary.

For liquid mercury to enter a cylindrical capillary or slit-shaped pore and achieve a flat surface it is necessary to apply some pressure. The dependencies of the pressure required for the introduction of non-wetting liquid into the capillary on its radius (or the half-width of the pore), on the nature of the liquid (via the liquid surface tension σ) and the nature of the solid (via the contact angle θ) are expressed by the formula [3].

$$r_p = -\frac{2\sigma \cos \theta}{\Delta P}, \tag{1}$$

where r_p is the radius of the cylindrical pore, m; ΔP is the pressure difference in the gas and liquid phases, divided by the curved meniscus, Pa.

3 Results and Discussion

In terms of results, initially, the Pascal mercury porosimeter was calibrated and valid metrological characteristics of measuring the intruded mercury volume and pressure were established. In the former case, a first-order measurement standard for the mass unit was used (Mettler Toledo XP205 laboratory balance having special accuracy class I). In the latter case, second-order measurement standards for the pressure unit were used (LR-CAL LPC 200 pressure calibrator[1]). On the basis of the calibration results, the following valid metrological characteristics of measuring the volume and pressure of mercury were established using the Pascal mercury porosimeter:

- the absolute pressure measurement ranges from 0.1 to 400 kPa and from 0.1 to 400 MPa, with respective expanded measurement uncertainties of 0.1 and 0.2%;

[1]The measurement limits of the LPC-2-6 and LPC-2-4000 pressure sensors used with the calibrator are 600 kPa and 400 MPa, with permissible absolute pressure fiducial error limits of ±0.025% and ±0.1%, respectively.

- the mercury volume measurement ranges from 5 to 500 mm^3 and from 20 to 2000 mm^3 depending on the model of the dilatometer used, with the expanded uncertainty being set at the upper measurement range value of 1.0%.

The results of calibrating the Pascal mercury porosimeter were further used for estimating the measurement uncertainty of porosity characteristics due to characterisation.

The procedure for measuring porosity characteristics using the Pascal mercury porosimeter consisted of several stages: the heat treatment of the material under study; conducting the blank experiment (measuring an empty dilatometer having no sample) at the stations of low and high pressure; sample analysis at the low- and high-pressure stations. At the low-pressure station, mercury was introduced into a vacuumised vessel containing the material, in which pressure measured with a pressure sensor was continually increased. As the mercury penetrated into the pores, its level in the vessel decreased, which was evaluated according to capacitance measurements. The dependence of intruded mercury on pressure is shown in graph form on the basis of the described experiment.

For the preliminary studies, the heat treatment of the material consisted of the following steps. Alumina-based ceramic membranes were dried prior to conducting measurements at a temperature of 105 °C for 3 h, while granular aluminium oxide was dried at a temperature of 150 °C for 3 h (desiccation to constant mass being the criterion for selecting the drying parameters). After drying, the sample intakes of the substances under analysis were cooled in a desiccator to ambient temperature; the density was measured five times using a gas pycnometer Pycnomatic ATC at the temperature of the measuring chamber (20.0 °C).

In the course of the blank experiment, the following operations were performed: (1) the height of the dilatometer cone was measured using a pair of callipers; (2) CD3/P parameters for the SOLID EVO software of the dilatometer were established: distance between electrodes set to 5 mm, dilatometer radius set to 1.5 mm; (3) an empty dilatometer was assembled and weighed with a laboratory balance; (4) the dilatometer was taken apart and silicone grease was thoroughly applied to the close-fitting parts; (5) the dilatometer was assembled, carefully rotating its close-fitting parts for uniform distribution of silicone grease, and weighed again; (6) an electrode was placed on the dilatometer and the construction was positioned on the Pascal 140 low pressure port, and the process of measuring the blank was launched at the pressure increase rate of 6–400 kPa and then lowering the pressure to 100 kPa in automatic mode[2]; (7) after conducting the measurements and filling the dilatometer with mercury, it was weighed again; (8) then the dilatometer construction was placed on the Pascal 440 high pressure port, and the process of measuring the blank was launched at pressure increase rate 6–400 MPa and the pressure subsequently lowered to 0.100 MPa in automatic mode.

In order to measure the material, analogous operations were carried out, with the difference being that the material under study was placed into the dilatometer

[2]The optimal rate was determined via preliminary studies, the minimum standard deviation of results of measuring the dominant pore size being the optimality criterion.

and its mass was recorded using the balance (special accuracy class I). The optimal mass of the sample intake was selected in such a way as to ensure specific-volume measurements of the intruded mercury at maximum pressure for over the half of the volumetric range. Measurements were conducted only at the low-pressure station (up to 400 kPa) for the materials having pore sizes of ~9000 and ~60,000 nm. For the materials having pores sizes of ~10 and ~2000 nm, the high-pressure station was used. Typical results for mercury intrusion and extrusion curves are shown in Fig. 1 alongside the pore size distribution curves for the materials under study.

4 Processing Mercury Porosimetry Measurement Results and Estimating Measurement Uncertainty Associated with Characterisation

The algorithm for calculating porosity characteristics using the SOLID EVO software of the Pascal mercury porosimeter is as follows: upon obtaining the mercury intrusion and extrusion curve, further corrections to pressure and volume readings are introduced.

For evaluating hydrostatic pressure H_2, MPa, the following equation is used at the starting point of the experiment

$$H_2 = \left(H_1 + \frac{V_{fil}}{\pi r_d^2} \right) \cdot 0.00013332, \tag{2}$$

where H_1 is the height (distance between the foundation, the dilatometer external cone connection point, and the base of the tube), mm;
V_{fil} is the mercury volume in the tube after filling the dilatometer under vacuum, mm^3;
r_d is the dilatometer radius, mm.
0.00013332 is the coefficient for converting the unit of hydrostatic pressure from mm Hg to MPa.

Over the course of the experiment, the pressure starts to increase. The mercury is gradually intruded into the material, with the result that the mercury level decreases in the capillary by the corresponding measured volume of the intruded mercury and lower hydrostatic pressure is observed.

$$H_i^{cor} = \left(\frac{V_{fil} - V_i}{\pi r_d^2} + H_1 \right) \cdot 0.00013332, \tag{3}$$

V_i is the volume of the mercury intruded into the material at the ith point of the mercury intrusion curve, mm^3.

Accordingly, the corrected pressure $\left(P_i^{cor}\right)$ is as follows

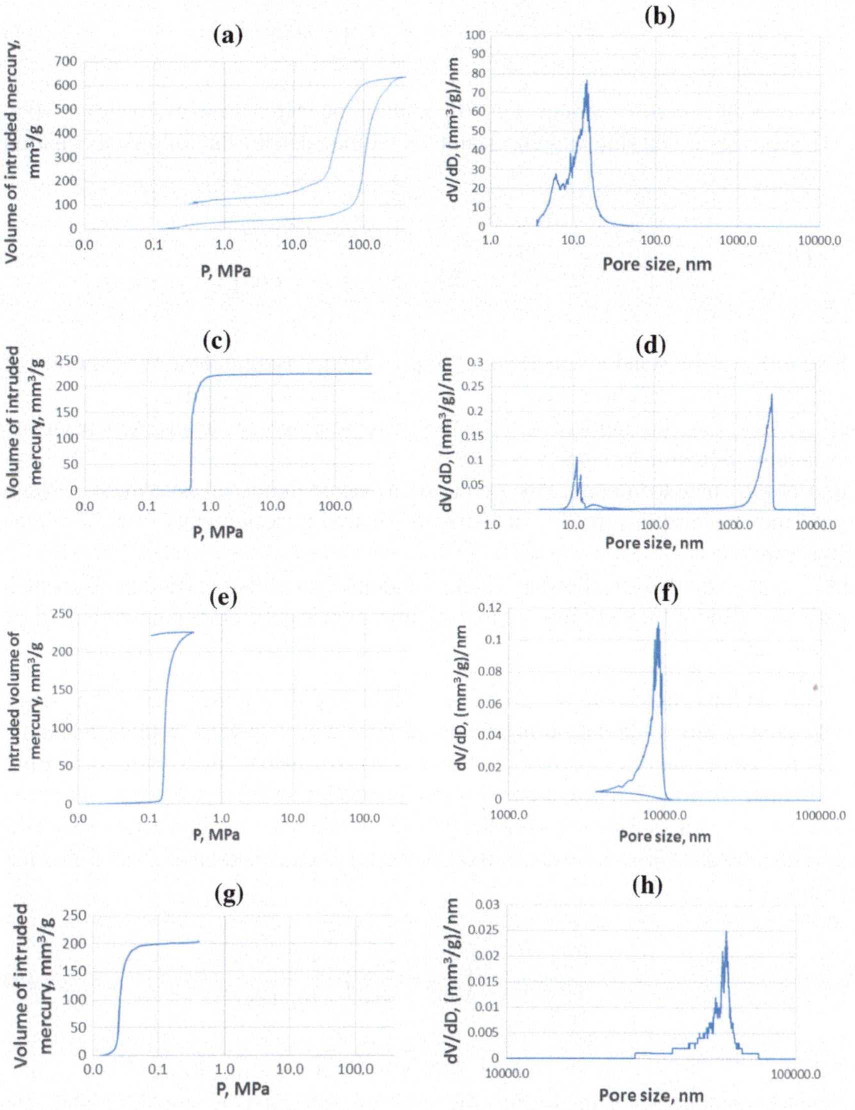

Fig. 1 Intruded mercury volume dependencies and corresponding pore size distribution curves: **a**, **b** granular alumina with pores sizes of ~10 nm, **c**, **d** macroporous ceramic membrane with pore sizes of ~2,000 nm, **e**, **f** macroporous ceramic membrane with pore sizes of ~9,000 nm, **g**, **h** macroporous ceramic membrane with pore sizes of ~60,000 nm)

$$P_i^{cor} = P_i + \left(\frac{V_{fil} - V_i}{\pi r_d^2} + H_1\right) \cdot 0.00013332, \quad (4)$$

where P_i is the pressure reading at the ith point of the mercury intrusion curve, MPa.

The corresponding standard uncertainty is estimated using the following equation

$$u\left(P_i^{cor}\right) = \sqrt{\begin{array}{l} u^2(P_i) + \left(\frac{0.00013332}{\pi r_d^2}\right)^2 \left(u^2\left(V_{fil}\right) + u^2(V_i)\right) \\ + \left(-\frac{2 \cdot 0.00013332\left(V_{fil} - V_i\right)}{\pi r_d^3}\right)^2 u^2(r_d) + (0.00013332)^2 u^2(H_1) \end{array}}, \quad (5)$$

where $u(P_i)$ is the standard uncertainty of pressure measurement at the ith point of the mercury intrusion curve using the built-in pressure sensors, MPa;
$u\left(V_{fil}\right)$, $u(V_i)$ are the standard uncertainties of mercury volume measurements using the mercury porosimeter, mm^3;
$u\left(r_d\right)$ is the standard uncertainty of the dilatometer radius measurements, which equals the standard uncertainty of linear dimension measurements using a microscope, mm;
$u\left(H_1\right)$ is the standard uncertainty of the dilatometer radius measurements, which equals the standard uncertainty of linear dimension measurements using a pair of callipers, mm.

Since a blank run is completed prior to conducting measurements,[3] and mercury volumes obtained when measuring the blank are then subtracted from each sample measurement result to compensate for the effects associated with the mercury heating during the experiment, hydrostatic pressure corrections were introduced both at the blank run, as well as for each sample analysis. To evaluate the correction based on the volume and its uncertainty, a correction factor is calculated using the following equation

$$k = \frac{V_{Hg}^{sample}}{V_{Hg}^{blank}} = \left(\frac{m_3^b - m_2^b}{m_4^s - m_3^s}\right) \cdot \frac{\rho_{Hg}(T_s)}{\rho_{Hg}(T_b)}, \quad (6)$$

where V_{Hg}^{sample}, V_{Hg}^{blank} are mercury volumes in the tube after filling the dilatometer under vacuum when measuring the material and carrying out the blank run, respectively, cm^3;

m_1^b is the dilatometer weight when measuring the blank, g;
m_2^b is the weight of the dilatometer and silicone grease when measuring the blank, g;
m_3^b is the weight of the dilatometer, silicone grease and mercury when measuring the blank, g;
m_1^s is the weight of the dilatometer when measuring the sample, g;
m_2^s is the weight of the dilatometer and sample when measuring the sample, g;

[3]Measurements are carried out for the empty dilatometer without a sample.

m_3^s is the weight of the dilatometer, sample and silicone grease when measuring the sample, g;
m_4^s is the mass of the dilatometer, sample, silicone grease and mercury when measuring the sample, g;
$\rho_{Hg}(T_s)$, $\rho_{Hg}(T_b)$ are the mercury densities when measuring the sample temperature and the blank temperature, respectively, g/cm^3, which are calculated based on the reference data of the SOLID software.

Then the blank experiment measurement results were corrected on the basis of the inequality of the mercury volume used in the blank experiment and the sample analysis. To do this, all the volumes of the blank experiment were initially multiplied by the correction factor

$$V_i^{blankcor} = V_i^{blank} \cdot k = V_i^{blank} \left(\frac{m_3^b - m_2^b}{m_4^s - m_3^s} \right) \cdot \frac{\rho_{Hg}(T_s)}{\rho_{Hg}(T_b)}, \tag{7}$$

where V_i^{blank} is the result of the mercury volume measurement at the ith intrusion curve point when conducting measurements for the blank experiment, mm^3.

The standard measurement uncertainty of corrected mercury volume of the blank run was calculated using the formula

$$u\left(V_i^{blankcor}\right) = \sqrt{\begin{array}{l} \left(\left(\frac{m_3^b - m_2^b}{m_4^s - m_3^s}\right) \cdot \frac{\rho_{Hg}(T_s)}{\rho_{Hg}(T_b)}\right)^2 u^2\left(V_i^{blank}\right) + \left(\frac{V_i^{blank}\rho_{Hg}(T_s)}{\rho_{Hg}(T_b)\left(m_4^s - m_3^s\right)}\right)^2 u^2\left(m_3^b\right) \\ + \left(-\frac{V_i^{blank}\rho_{Hg}(T_s)}{\rho_{Hg}(T_b)\left(m_4^s - m_3^s\right)}\right)^2 u^2\left(m_2^b\right) + \left(\frac{V_i^{blank}\rho_{Hg}(T_s)\left(m_3^b - m_2^b\right)}{\rho_{Hg}(T_b)\left(m_4^s - m_3^s\right)^2}\right)^2 u^2\left(m_4^s\right) \\ + \left(\frac{-V_i^{blank}\rho_{Hg}(T_s)\left(m_3^b - m_2^b\right)}{\rho_{Hg}(T_b)\left(m_4^s - m_3^s\right)^2}\right)^2 u^2\left(m_3^s\right) + \left(\frac{V_i^{blank}\left(m_3^b - m_2^b\right)}{\rho_{Hg}(T_b)\left(m_4^s - m_3^s\right)^2}\right)^2 \\ u^2\left(\rho_{Hg}(T_s)\right)\left(\frac{-V_i^{blank}\rho_{Hg}(T_s)\left(m_3^b - m_2^b\right)}{\rho_{Hg}^2(T_b)\left(m_4^s - m_3^s\right)^2}\right)^2 u^2\left(\rho_{Hg}(T_b)\right) \end{array}}, \tag{8}$$

where $u\left(\rho_{Hg}(T_b)\right) = u\left(\rho_{Hg}(T_s)\right) = u\left(\rho_{Hg}\right) = 0.0012$ g/cm^3 is the standard uncertainty of mercury density measurements, g/cm^3 (the estimate being made in comparison with the data presented in [8]);
$u\left(m_3^b\right) = u\left(m_2^b\right) = u\left(m_4^s\right) = u\left(m_3^s\right) = u(m)$ is the standard uncertainty of mass measurements, g;
$u\left(V_i^{blank}\right)$ is the standard uncertainty of mercury volume measurements obtained using the mercury porosimeter at the i-th intrusion curve point when conducting measurements for the blank experiment, mm^3;

To subtract the blank run results from the results of the experiment where a sample was used, preliminary data interpolation of the blank is conducted in relation to the corresponding pressures when measuring the material under analysis, since pressures in the blank run and in the experiment analysing the material differ. For conducting the data interpolation of the blank using the SOLID software, the following equation is applied

$$\begin{aligned}
\widehat{V}_c^{blankcor} &= P_c\left(\frac{V_{i+1}^{blankcor}-V_i^{blankcor}}{P_{i+1}^{blankcor}-P_i^{blankcor}}\right)+V_i^{blankcor} \\
&\quad - P_i^{blankcor}\left(\frac{V_{i+1}^{blankcor}-V_i^{blankcor}}{P_{i+1}^{blankcor}-P_i^{blankcor}}\right) \\
&= \left(\frac{V_{i+1}^{blankcor}-V_i^{blankcor}}{P_{i+1}^{blankcor}-P_i^{blankcor}}\right)\left(P_c-P_i^{blankcor}\right) \\
&\quad + V_i^{blankcor},
\end{aligned} \tag{9}$$

where P_c is the pressure in analysing the material, corrected for hydrostatic pressure, MPa. P_c is in a range $P_i^{blonkor} < P_c < P_{i+1}^{blankcol}$, MPa; $V_{i+1}^{blankcor}$, $V_i^{blankcor}$, $u\left(V_{i+1}^{blankcor}\right)$, $u\left(V_i^{blankcor}\right)$ are the corrected volumes of the blank experiment and the standard uncertainties, which correspond to the ith and $i+1$th points of the mercury intrusion curve, respectively, mm^3;
$P_{i+1}^{blankcor}$, $P_i^{blankcor}$, $u\left(P_{i+1}^{blankcor}\right)$, $u\left(P_i^{blankcor}\right)$ are the corrected pressures of the blank experiment and the standard uncertainties, which correspond to ith and $i+1$th points of the mercury intrusion curve, MPa;
$\widehat{V}_c^{blankcor}$ is the interpolated volume of the blank experiment, which is subtracted from the volume obtained during the sample analysis with the corresponding pressure P_c, mm^3.

The standard uncertainty of the interpolated mercury volume in the blank experiment was calculated using the formula

$$\begin{aligned}
&u\left(\widehat{V}_c^{blankcor}\right) \\
&= \sqrt{\begin{aligned}
&\left(\frac{P_i^{blankcor}}{P_i^{blankcor}-P_{i+1}^{blankcor}}-\frac{P_c}{P_i^{blankcor}-P_{i+1}^{blankcor}}\right)^2 u^2\left(V_{i+1}^{blankcor}\right) \\
&+\left(\frac{P_c}{P_i^{blankcor}-P_{i+1}^{blankcor}}-\frac{P_i^{blankcor}}{P_i^{blankcor}-P_{i+1}^{blankcor}}+1\right)^2 u^2\left(V_i^{blankcor}\right) \\
&+\left(\frac{P_i^{blankcor}\left(V_i^{blankcor}-V_{i+1}^{blankcor}\right)}{\left(P_i^{blankcor}-P_{i+1}^{blankcor}\right)^2}-\frac{\left(V_i^{blankcor}-V_{i+1}^{blankcor}\right)}{P_i^{blankcor}-P_{i+1}^{blankcor}}-\frac{P_c\left(V_i^{blankcor}-V_{i+1}^{blankcor}\right)}{\left(P_i^{blankcor}-P_{i+1}^{blankcor}\right)^2}\right)^2 u^2\left(P_i^{blankcor}\right) \\
&+\left(\frac{P_c\left(V_i^{blankcor}-V_{i+1}^{blankcor}\right)}{\left(P_i^{blankcor}-P_{i+1}^{blankcor}\right)^2}-\frac{P_i^{blankcor}\left(V_i^{blankcor}-V_{i+1}^{blankcor}\right)}{P_i^{blankcor}-P_{i+1}^{blankcor}}\right)^2 u^2\left(P_{i+1}^{blankcor}\right)
\end{aligned}},
\end{aligned} \tag{10}$$

The specific pore volume and corresponding standard uncertainty when analysing the material are calculated according to the following equations

$$\widehat{V}_i^s = \frac{V_i^s - \widehat{V}_c^{blankcor}}{m_2^s - m_1^s}, \quad (11)$$

$$u\left(\widehat{V}_i^s\right) = \sqrt{\left(\frac{1}{m_2^s - m_1^s}\right)^2 \left(u^2(V_i^s) + u^2\left(\widehat{V}_c^{blankcor}\right)\right) + \left(\frac{\widehat{V}_c^{blankcor} - V_i^s}{(m_1^s - m_2^s)^2}\right)^2 \left(u^2(m_2^s) + u^2(m_1^s)\right)}, \quad (12)$$

where V_i^s is the result of measuring the intruded mercury volume at the ith point when analysing the sample, mm^3.

The pore size (width or diameter) is calculated using the Washburn equation

$$D = -\frac{4 \cdot \sigma \cdot \cos\theta}{P}, \quad (13)$$

where P is the corrected pressure value when analysing the material, Pa
σ is the mercury surface tension, N/m;
θ is the wetting contact angle, rad.

In line with [9], the constant values of the mercury surface tension (0.484 N/m) and the wetting contact angle (140°) are to be applied for mercury, regardless of the chemical nature of the material.

The type B standard uncertainty of pore size is calculated using the following equation

$$u_B(D) = \sqrt{\left(\frac{4 \cdot \sigma \cdot \cos\theta}{P^2}\right) u^2(P) + \left(-\frac{4 \cdot \cos\theta}{P}\right)^2 u^2(\sigma) + \left(\frac{4 \cdot \sin\theta}{P}\right)^2 u^2(\theta)}, \quad (14)$$

where $u(P)$ is the standard uncertainty of the corrected pressure value when analysing the sample, Pa;
$u(\sigma)$ is the standard uncertainty of mercury surface tension, N/m (in the calculations assumed to be equal to zero);
$u(\theta)$ is the standard uncertainty of the wetting contact angle of the sample surface, rad (in the calculations assumed to be equal to zero).

The total standard uncertainty was calculated by the following equation

$$u_c(D) = \sqrt{u_A^2(D) + u_B^2(D)}, \quad (15)$$

where $u_A(D) = \sqrt{\frac{\sum (D_i - \bar{D})^2}{n(n-1)}}$ is the standard uncertainty of type A.

The expanded uncertainty at k = 2 and P = 0.95 was calculated using the following equation

$$U = ku_c(D), \tag{16}$$

The validation of the procedure for reproducing porosity characteristics and of the uncertainty calculation algorithm was performed using foreign CRMs, ERM-FD122 and BAM-P125, produced by BAM (Germany) (Tables 1 and 2). The results showed a satisfactory agreement between the measurement results obtained using the reference system and RM certified values taking into account the expanded uncertainty associated with characterisation, evaluated in line with the developed algorithm using the mercury porosimetry method. The measurement result quality criterion E_n was calculated using the following formula.

$$E_n = \frac{x - X}{\sqrt{U^2(x) + U^2(X)}}, \tag{17}$$

where x, $U(x)$ are the measurement result and corresponding expanded uncertainty of porosity characteristics;
X, $U(X)$ are the certified value and corresponding expanded uncertainty for certified reference materials.

Table 1 Data summary on measuring porosity characteristics of the CRM ERM-FD122

Characteristic designation	V (P = 100 MPa), mm^3/g	V (P = 195 MPa) mm^3/g	V (P = 200 Mpa), mm^3/g	V (P = 395 MPa) mm^3/g	Average pore width, nm	Most frequent pore width, nm
Average value	922.8	925.3	925.3	925.7	140.4	141.8
u(A)	5.2	5.8	5.8	5.7	2.6	1.7
u(B)	12.0	12.1	12.6	12.1	2.9	2.9
uc	13.1	13.4	13.9	13.4	3.9	3.4
U (k = 2, P = 0.95)	26.1	26.8	27.7	26.7	7.9	6.7
Uo (k = 2, P = 0.95), %	2.8	2.9	3.0	2.9	5.7	4.8
RM certified value	919.7	922.5	922.6	924.4	139.0	140.2
Expanded uncertainty of the certified value	16.8	17.5	17.5	17.2	3.7	3.9
E_n	0.1	0.1	0.1	0.0	0.2	0.2
Conclusion	+	+	+	+	+	+

Note u(A) is the type A standard uncertainty; u(B) is the type B standard uncertainty; u_C is the combined standard uncertainty; Uo (k = 2, r = 0.95) is the relative expanded uncertainty at k = 2; E_n is the quality criterion for the measurement results; V (P) is the specific pore volume at pressure P

Table 2 Data summary on measuring porosity characteristics of the CRM BAM-P125

Characteristic designation	Y2, MPa	Y3, MPa	P_{50}, MPa	Y1 mm^3/g	Average pore width, nm
Average:	0.2516	0.1289	0.2420	213.7	6,032.4
u(A)	0.0050	0.0026	0.0048	2	60
u(B)	0.0005	0.0006	0.0002	7.0	12.1
uc	0.0051	0.0026	0.0048	7.3	61.5
U(k = 2, P = 0.95)	0.0101	0.0053	0.0097	14.6	123.0
Uo (k = 2, P = 0.95), %	4.0	4.1	4.0	6.8	2.0
RM certified value	0.2646	0.1366	0.2554	207.9	5,796
Expanded uncertainty of the RM certified value	0.0135	0.0179	0.00940	11.0	215
E_n	−0.8	−0.4	−1.0	0.3	1.0
Conclusion	+	+	+	+	+

Note u (A) is the type A standard uncertainty of; u(B) is the type B standard uncertainty; u_C is the combined standard uncertainty; Uo (k = 2, r = 0.95) is the relative expanded uncertainty at k = 2; E_n is the quality criterion for measurement results; Y1 is the specific pore volume at 0.88 MPa (specific saturation volume); Y2 is the pressure corresponding to 57.5% of the specific saturation volume; Y3 is the pressure difference corresponding to 87.5 and 25% of the specific saturation volume; P_{50} is the pressure corresponding to 50% of the specific saturation volume

If $|E_n| \leq 1, 0$, the results are characterized as satisfactory and no further action is required, whereas $|E_n| > 1, 0$ characterizes results as unsatisfactory necessitating some action (Table 1).

The homogeneity of the RMs in question was studied in line with [10]. Granular alumina was packaged in separate glass bottles and several sample intakes were selected from each of them and measured for porosity characteristics using the Pascal mercury porosimeter. The RM material consisted of macroporous ceramic membranes in the form of discs (8–12 mm thick, 50–60 mm in diameter). Each disc was sawn into a large number of small separate sample intakes, weighing about (0.5–1.5) g, suitable for carrying out one measurement. At least two sample intakes were selected from each disc and were measured using the Pascal mercury porosimeter. The granular alumina stability in terms of porosity characteristics was studied using the gas adsorption method, while the stability of various microporous membranes for a period of 3 years was confirmed by research of BAM [5]. Thus, the standard uncertainty due to instability was seen as irrelevant for at least 24 months. The preliminary data summary on the results of estimating RM metrological characteristics is shown in Table 3. The analysis of the data presented in Table 3 shows contributions arising from the type A standard uncertainty, the type B standard uncertainty, which are caused by the characterisation, as well as the standard uncertainty due to heterogeneity of the RM material. In accordance with these contribution estimations, which are presented in Table 3, it can be rather difficult to single out the dominant

Table 3 Data summary on the metrological characteristics of the RMs under development

RM matrix	Quantity designation	Certified value	RM metrological characteristics				
			uAo	*uBo*	*uho*	uco*(A)*	Uo*(A)*
Granular alumina (pore size 10 nm)	Specific pore volume, mm^3/g	624.2	0.27	1.14	0.6	1.3	2.6
	Most frequent pore diameter, nm	14.19	1.21	0.30	2.4	2.7	5.4
	Pore median, nm	13.81	0.87	0.31	1.7	2.0	3.9
Macroporous ceramic membrane No. 1 (pore size 9,000 nm)	Specific pore volume, mm^3/g	218.7	0.69	1.83	1.31	2.4	4.7
	Most frequent pore diameter, nm	8,668	0.85	0.13	1.62	1.8	3.7
	Pore median, nm	8,315	0.76	0.13	1.45	1.6	3.3
Macroporous ceramic membrane No. 2 (pore size 60,000 nm)	Specific pore volume, mm^3/g	197.7	0.31	1.00	1.78	2.1	4.1
	Most frequent pore diameter, nm	57,985	0.71	0.81	0.29	1.1	2.2
	Pore median, nm	56,368	0.63	0.83	1.07	1.5	3.0
Macroporous ceramic membrane No. 3 (pore size 2,000 nm)	Specific pore volume, mm^3/g	224.7	0.43	2.31	1.4	2.7	5.5
	Most frequent pore diameter, nm	1,847	0.90	1.80	1.6	2.6	5.2
	Pore median, nm	2,627	2.03	1.40	2.5	3.5	7.0

Notes *uAo* is the type A relative standard uncertainty; *uBo* is the type B relative standard uncertainty; *uho* is the relative standard uncertainty due to heterogeneity; *uco (A)* is the relative combined standard uncertainty; *Uo (A)* (k = 2, P = 0.95) is the relative expanded uncertainty of the RM certified value at k = 2 and P = 0.95

contribution, since depending on the measurand and the matrix of the RM material under study, the ratio between contributions changes significantly.

Thus, four RM types, which are not inferior to foreign analogues [5, 6], have been developed. The pores sizes of these RMs range from 10 nm up to 60,000 nm, with relative expanded uncertainty of certified values from 2.2% to 7.0%, respectively. Further accuracy improvement when reproducing specific volume and pore size using the reference system is possible by means of accurate calibration of the built-in pressure sensors and capacitive sensors for measuring the mercury volume, searching for more homogeneous porous materials suitable for creating RMs, as well as by conducting a large number of duplicate measurements.

5 Conclusions

Thus, in carrying out this work, a procedure for reproducing the size and specific pore volume of macroporous ceramic membranes and granular alumina, as well as a corresponding procedure for calculating uncertainty due to characterisation, were developed. The certified value traceability of the RMs under development is provided by the application of the following measurement standards for calibrating the mercury porosimeter: the second-order measurement standards for the pressure unit and the first-order measurement standard for the mass unit. When assessing the metrological characteristics of RMs, the standard uncertainty, associated with the characterisation and heterogeneity of the materials under study (RM candidates), was estimated. The standard uncertainty due to instability was disregarded due to the high stability of the alumina matrix. According to the results of the studies, four types of alumina-based porosity RM were developed, with pore sizes of ~10 nm; ~2,000 nm; ~9,000 nm; ~60,000 nm. The RMs with pore sizes of ~10 nm; ~2,000 nm are intended for use in controlling the accuracy of mercury porosimeters operating at high pressures, while the RMs with pore sizes of ~9,000 nm; ~60,000 nm are for controlling accuracy of mercury porosimeters operating at low pressures. The mercury porosimetry RMs are being developed for the first time in Russia and are primarily intended for verifying/calibrating and testing mercury porosimeters for type approval. The developed RMs are different from their foreign counterparts, which are produced by BAM (Germany) and NIST (USA) metrological institutes. The standard uncertainty due to characterisation of the RMs in question was estimated, not by the method of interlaboratory experiment, but by a full analysis of the constraint equations taking into account all sources of uncertainty included in the equations on the basis of the Guide to the Expression of Uncertainty in Measurement. The conducted studies allowed the RM certified values to be traceable at this stage to the GET 3-2008 State Primary Measurement Standard for the mass unit (kilogram), the GET 101-2011 State Primary Measurement Standard for the pressure unit in the field of absolute pressure in the range 1×10^{-1}–7×10^{5} Pa, the GET 43-2013 State Primary Measurement Standard for the pressure unit in the range 10–1600 MPa and the effective area of the piston-cylinder unit of dead weight testers in the range 0.05–1 cm^2. In addition, a reference material having a large pore size of ~60,000 nm, which currently has no analogues elsewhere in the world, was characterised. The development of type approved RMs will ensure metrological support of porosity measuring instruments for solid substances and materials based on the mercury porosimetry method in various industrial sectors. Thus, the metrological independence of the Russian

Federation will be ensured and import substitution of expensive foreign reference materials carried out.

Conflict of Interest One of the authors is a member of the Editorial Board of these proceedings.

References

1. Karnaukhov AP (1999) Adsorption. Texture of dispersed and porous materials. Nauka, SB RAS, Novosibirsk (in Russ.)
2. Fenelonov VB (2004) Introduction to the basics of adsorption and texturology. Publ. House of SB RAS, Novosibirsk (in Russ.)
3. Grek S, Sing K (1984) Absorption, specific surface, porosity. Moscow, Mir, (in Russ.)
4. GOST ISO Guide 35-2015 Reference materials. General and statistical principles for certification
5. BAM official website (2018) Reference Materials. Certificates and reports. https://rrr.bam.de/RRR/Navigation/EN/Reference-Materials/RM-Certificates-reports/Porous-RM/porous-rm.html. Accessed 3 Oct 2018
6. NIST (2011) A certificate of analysis on the SRM 1918. Mercury porosimeter intrusion standard. https://www-s.nist.gov/srmors/certificates/1918.pdf. Accessed 03 Oct 2018
7. Order of the Federal Agency for Technical Regulation and Metrology of January 29, 2015 No. 128. On approval of the national primary standard for the units of specific adsorption of gases, specific surface area, specific volume and pore size of solid substances and materials (in Russ.)
8. GSSSD 3-77 Mercury. Mercury density and the coefficient of thermal expansion at atmospheric pressure and temperatures from 0 to 350 °C. Reference data tables
9. ISO 15901-1:2016 Evaluation of pore size distribution and porosity of solid materials by mercury porosimetry and gas adsorption—Part 1: Mercury porosimetry
10. Sobina EP (2016) Development of a Certified Standard Reference Sample of Nanoporous Aluminum Oxide. Meas Tech 8:68–72 (in Russ.)

Development of a Reference Material for the Composition of 0.1M Potassium Dichromate Solution. Certified Reference Material GSO 10992-2017

Alexander Yu. Shimolin and Alena V. Sobina

Abstract This paper describes key stages in the development and investigation of a high-precision certified reference material (CRM) for the composition of potassium dichromate solution. The proposed reference material (RM) exhibits metrological traceability to the units of the International System of Units (SI). Not requiring additional preparation before use, the CRM is convenient for use as a titrant in redox titration. The metrological characteristics of the developed CRM were determined by the direct primary method of coulometric titration using the State Primary Standard GET 176. A procedure for measuring the molar concentration of potassium dichromate is described, along with the equipment used. A comprehensive uncertainty budget for the measurements is assembled. The interval of permissible certified values of the molar concentration of potassium dichromate in the CRM under study is set to vary from 0.098 to 0.102 mol/dm^3. The relative expanded uncertainty (k = 2) of the certified value of potassium dichromate molar concentration is calculated to be 0.024%, with the relative standard uncertainties caused by long-term and short-term instability each being equal to 0.0044%. The shelf life of the developed CRM is found to be 1 year, provided that storage conditions are ensured. The developed CRM is included into the State Register of Certified Reference Materials under the number GSO 10992-2017. The CRM can be used for transferring the mass (molar) concentration unit of a component to RMs and chemical reagents via oxidation-reduction reactions; the verification and calibration of measuring instruments (MI), as well as control of metrological characteristics during MI trials including type approval; MI graduation; validation of measurement procedures; control of measurement results when determining the mass (molar) fraction and mass (molar) concentration of components in liquid and solid substances and materials.

Keywords Coulometry · Coulometric titration · Titrimetry · State primary standard · GET 176 · Potassium dichromate solution · Reference material · Certified value uncertainty · Potassium dichromate · Traceability

---

A. Yu. Shimolin (✉) · A. V. Sobina
Ural Scientific Research Institute for Metrology (FGUP UNIIM), Ekaterinburg, Russia
e-mail: alex-shimolin@uniim.ru

S. Medvedevskikh et al. (eds.), *Reference Materials in Measurement and Technology*,
https://doi.org/10.1007/978-3-030-32534-3_11

Glossary of Abbreviations

GET 176 State primary standard of units of mass (molar) fraction and mass (molar) concentration of components in liquid and solid substances and materials based on coulometry GET 176-2017;
RM Reference material;
CRM Certified reference material;
MI Measuring instrument;
BIPM Bureau International des Poids et Mesures—International Bureau of Weights and Measures;
CMC Calibration and measurement capabilities.

1 Introduction

Due to both high stability, including that in solutions, and strong oxidative properties, potassium dichromate is widely used as a reference material (RM) in titrimetry for determining reducing agents.

For the purposes of achieving the highest possible accuracy and ensuring metrological traceability to the SI, potassium dichromate-based titrants are prepared from potassium dichromate RMs certified by the primary method of coulometric titration [1] in national measurement institutes [2]. Thus, in Russia, GSO 2215-81[1] of potassium dichromate certified using GET 176[2] via the method of coulometric titration [3] is broadly applied. The accuracy of mass fraction measurement results of the main component in potassium dichromate has been confirmed by bilateral [3] and international comparisons [5]. This information is included into the database of Calibrating and Measuring Capabilities (CMC)[3] of the International Bureau of Weights and Measures (BIPM).

However, further stages in the calibration chain from the GSO 2215-81 to reference materials, chemical reagents and their solutions contain a number of factors, which might affect the accuracy of measurement results in the case when the laboratory staff lacks the necessary skills or experience. These factors include such operations as drying of solid substance to constant weight, preparation of titrants of a required concentration, calibration of measurement glassware, etc. The influence of these factors in the verification of measurement results can be eliminated by using ready titrant solutions. Thus, there are available potassium dichromate solutions of

[1]State standard reference material of the first class (CRM) 2215—potassium dichromate RM. Available at Rosstandart: https://fgis.gost.ru/fundmetrology/registry/19/items/393979.

[2]GET 176-2017 State primary standard for units of mass (molar) fraction and mass (molar) concentration of a component in liquid and solid substances and materials based on coulometry. Available at Federal State Unitary Enterprise UNIIM: https://uniim.ru/gauge_fgup_uniim/

[3]Calibration and Measurement Capabilities. Amount of substance, High purity chemicals, Russian Federation. Available at: https://kcdb.bipm.org/appendixC/QM/RU/QM_RU_1.pdf.

a certain concentration suitable for classical volumetric titration and traceable to the potassium dichromate certified reference materials produced by national metrological institutes, such as Potassium Dichromate solution produced by Carl Roth GmbH[4] and Sigma-Aldrich.[5] These solutions are convenient to use; however, they have the expanded uncertainty (k = 2) of certified values of (0.2–0.3)%, which, in some cases, exceeds the target uncertainty of measurement results by the titration method.

In Russian practice, potassium dichromate titration standards produced in glass ampoules and containing potassium dichromate in accurately weighted portions that require dilution with distilled water in 1 L capacity measuring flasks are widely used. Examples include the Potassium Dichromate Titration Standard (Technical Specification) produced by the Himreactiv, LLC[6] and the Ural Plant of Chemical Products.[7] Following the preparation of a solution, a potassium dichromate solution with a titre of (0.100 ± 0.001) mol-equivalent/l is obtained. In the 2000s, reference materials of titrants being more accurate analogues of standard titres (relative uncertainty of potassium dichromate concentration in solution ±0.3%) were developed [6]. Prepared according to the described procedure, however, potassium dichromate solutions not only had inappropriate metrological characteristics for high-accuracy measurements, but also required additional preparation under laboratory conditions before use.

In order to provide laboratories with ready potassium dichromate titrant solutions exhibiting high metrological characteristics (with a higher accuracy than those described above) and traceability to SI units, a CRM for the composition of potassium dichromate solutions has been developed. The proposed CRM does not require additional preparation operations under laboratory conditions and is comparable in accuracy with the GSO 2215-81. The CRM for the composition of potassium dichromate solutions has become the second in the list of high-accuracy CRMs for the composition of pure chemical solutions after the hydrochloric acid solution [7] developed by the Ural Scientific Research Institute for Metrology (UNIIM). The CRMs have been certified by the coulometric titration method using GET 176.

2 Measurement Methodology

The CRM was developed using the State primary standard of units of mass (molar) fraction and mass (molar) concentration of components in liquid and solid substances

[4]CAS No. 7778-50-9 Potassium dichromate solution. Available at: https://www.carlroth.com/en/en/Chemicals/A-Z-Chemicals/P/Potassium-dichromate-solution-%280%2C1-N%29/Potassium-dichromate-solution/p/000000010000965500020023_en.

[5]Potassium dichromate solution. Available at: https://www.sigmaaldrich.com/catalog/product/mm/109118?lang=en&region=RU.

[6]Potassium dichromate standard titre (Technical Specification). Available at: http://himreactiv.com/catalog/20799/.

[7]Potassium dichromate standard titre. Available at: http://www.upcr.ru/catalog/gso_i_standart_titry/.

and materials based on coulometry GET 176. Assay tests of potassium dichromate were carried out by the method of coulometric titration of dichromate ions by bivalent iron ions electrogenerated on platinum cathode in a trivalent iron solution [8]. This process involved the following reactions:

$$Fe^{3+} + e \rightarrow Fe^{2+} \quad (cathode)$$

$$Cr^{6+} + 3Fe^{2+} \rightarrow Cr^{3+} + 3Fe^{3+} \quad (solution)$$

The mass of reacted potassium dichromate was determined by the electrical charge passed through the solution from the beginning to the end point of titration (EPT), according to the Faraday law

$$m = \frac{M \cdot Q}{z \cdot F}, \tag{1}$$

where

- m is the mass of the analyzed substance sample, g;
- M is the molecular mass of the analyzed substance, equal for potassium dichromate to 294.1818 g/mol [9];
- Q is the amount of electricity spent during the titrant generation, C;
- z is the number of electrons participating in the reaction (z= 6);
- F is the Faraday constant equal to 96485,33289 C/mol [10].

The main equation for determining the molar concentration of potassium dichromate using GET 176 can be represented as follows:

$$C_{mol} = \frac{1000 \cdot \rho \cdot U/R \cdot t}{z \cdot F \cdot m}, \tag{2}$$

where

- ρ is the density of the analyzed solution, g/ml;
- U is the voltage of generating current, V;
- R is the reference resistor value, ohm;
- t is the generation time, s.

3 Equipment and Reagents

Figure 1 displays an installation scheme for the coulometric titration method and the equipment used, which correspond to [11].

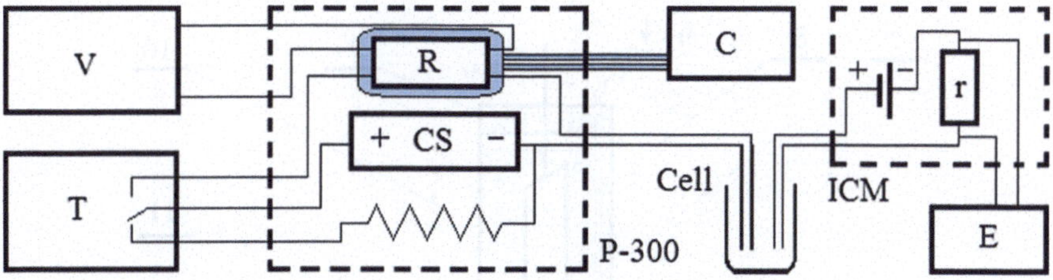

Fig. 1 Installation for the coulometric titration method on GET 176

Generating current from a highly-stable precision current source (CS) P-300 (UNIIM, Russia) was applied to a coulometric cell (Cell). The current was controlled by a voltage dip on the reference resistor (R) VFZP (10 Ω, Vishay Precision Group, USA) using a precision multimeter (V) 34470A (Keysight Technologies, USA) and a thermostat (C) NanoTherm3 (±0.1 °C, ThermoTek, USA). The current generation time was controlled by a frequency counter (T) 53131A (Agilent Technologies, USA) synchronized with the CS. Two platinum nets with a surface area of ~80 cm^2 were used as working and auxiliary electrodes. The system of biamperometric indication of the titration end point is presented in the form of an indicator current module (ICM) (UNIIM, Russia), which includes a voltage source (±) and an auxiliary resistor (r) with a rating of 1 MΩ, as well as a liquid analyser (E) *Expert-001-1* (Econix-Expert, Russia) used as a voltmeter with high input resistance and two platinum needle electrodes.

The sample mass of potassium dichromate solution was determined using a Sartorius CP225D laboratory balance (Germany) taking into account air buoyancy correction [12].

A vertical two-compartment coulometric cell was used for measurements, which scheme is shown in Fig. 2.

The 0.4M solution of ammonium ferric (III) sulfate in the 2M solution of sulfuric acid (hereinafter referred to as the catholyte) was used as the working electrolyte. The 2M solution of sulfuric acid (hereinafter referred to as the anolyte) was used as the auxiliary chamber electrolyte. Spaces in the cathodic and anodic cameras are securely divided with an electro-permeable microporous glass frit.

Analytical grade potassium dichromate (specification according to GOST 4220-75 [13]) served as the material for CRM preparation. The recrystallized material was dissolved in deionized water with an electrolytic conductivity not exceeding 1×10^{-5} S/m to obtain 0.1M solution.

4 Measurement Procedure

The catholyte in the amount of 100 cm^3 was poured into the working chamber. The cover was tightly closed. The auxiliary chamber, electrodes, magnetic stir bar, gas

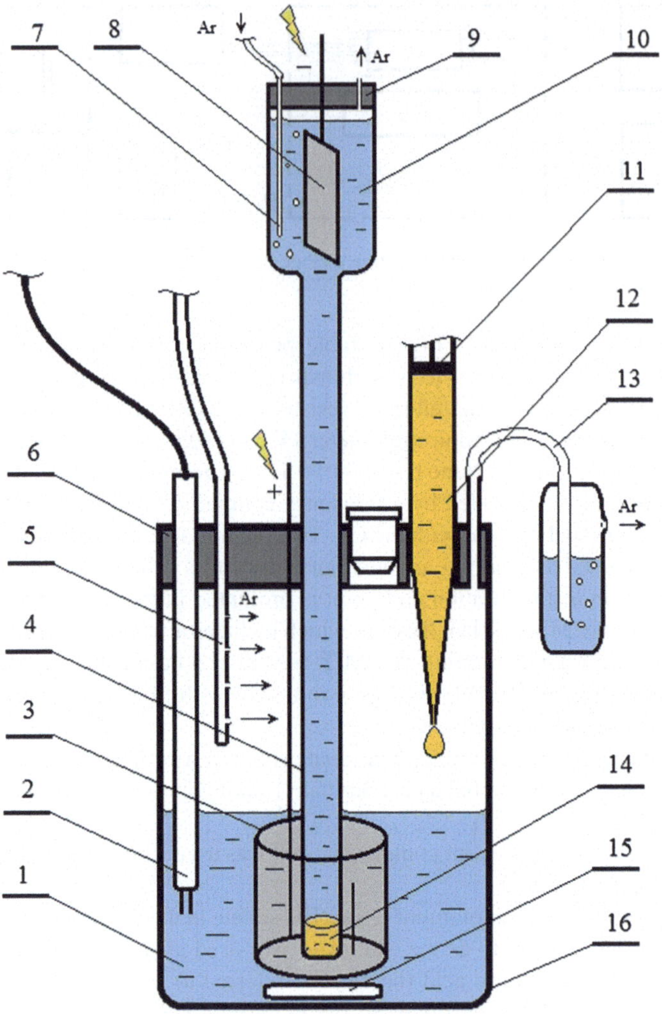

Fig. 2 The scheme of the coulometric cell 1—catholyte; 2—indicating electrodes; 3—generating electrode; 4—auxiliary chamber; 5—argon flow distributor; 6—cell cover; 7—auxiliary chamber bubbler; 8—auxiliary electrode; 9—auxiliary chamber cover; 10—anolyte; 11—injection system for a liquid sample with feedback; 12—liquid sample ($K_2Cr_2O_7$); 13—gas vent with a hydraulic lock; 14—membrane; 15—magnetic stir bar; 16—working chamber

vent with a hydraulic lock—were all placed into the cell according to the scheme depicted in Fig. 2, alongside with a bubbler. The latter is not shown in Fig. 2; it can be fixed in the place of the argon flow distributor. The cell was checked for tightness. The anolyte in the amount of 100 cm^3 was poured into the auxiliary chamber followed by the immersion of the auxiliary electrode and the auxiliary chamber bubbler. The magnetic stir bar was then activated. The cell was being deaerated with high-purity

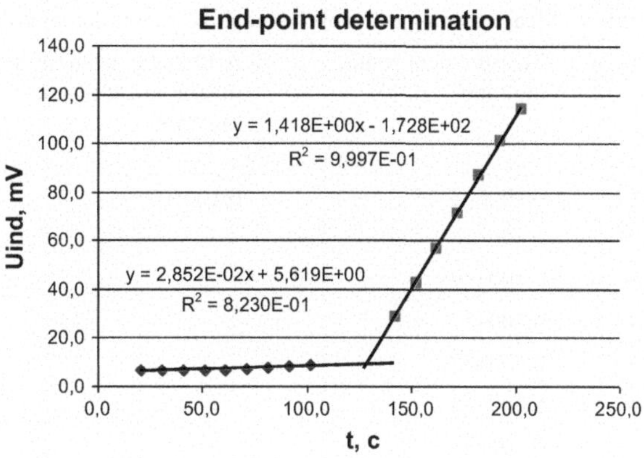

Fig. 3 A typical preliminary/final titration curve

argon during 1 h, with the gas flowing through the electrolyte. The bubbler was subsequently replaced with a gas-distributing pipe, with the gas being blown over the solution during all measurements (gas consumption was at least 100 cm^3/min).

Pre-generation was carried out by a 1 mA current with 10-s pulses. To this end, 0.1M potassium dichromate solution in the amount of 20 mg was introduced into the working chamber, which had been previously deaerated with argon. Charge was then being generated by 20-s pulses until the indicator signal changed. Afterwards, the titration was completed by at least six 4-s pulses. The values of the generating current voltage, current pulse length and indicating signal (U_{ind}) were recorded using a computer software. Using to the data obtained, linear dependences $U_{ind} = f(t)$ for the horizontal and sloping graph areas were plotted. The starting titration point was determined as the cross point of two linear functions. A typical preliminary titration curve is shown in Fig. 3.

The main part of charge generation was carried out by 160 mA current. 0.1M potassium dichromate solution in the approximate amount of 2.6 g was being continually introduced into the cell in a dropwise manner over the entire generation period of about 1,000 s. It should be noted that the generation time was calculated in such a way that 0.1% of potassium dichromate remained non-titrated. The sample mass was determined by the difference between two weighings, those made before and after sample introduction. After the completion of the main generation, the titration of excess potassium dichromate was started in a manner similar to pre-generation.

5 Processing of Measurement Results

The CRM certified value was estimated as the arithmetic mean of nine measurements of the molar concentration of potassium dichromate in the CRM. The respective data is presented in Table 1.

Table 1 Measurement results of the molar concentration of potassium dichromate

S. No.	Measurement result, mol/dm^3	Arithmetic mean of the measurement results, mol/dm^3	Relative standard deviation of the arithmetic mean of measurement results, %
1	0.100502	0.100509	0.0085
2	0.100516		
3	0.100489		
4	0.100477		
5	0.100489		
6	0.100488		
7	0.100538		
8	0.100545		
9	0.100536		

Measurement uncertainty was calculated according to [14]. The standard type A uncertainty was evaluated as a standard deviation of the arithmetic mean of potassium dichromate molar concentration measurement results using the formula

$$u_A = \sqrt{\frac{1}{n(n-1)} \sum_{i=1}^{n} \left(C_{mol,i} - \bar{C}_{mol}\right)^2}, \tag{3}$$

where

$C_{mol,i}$ is the ith measurement result of the output value, mol/dm^3

$\bar{C}_{mol}$ is the arithmetic mean of the output value in a series of n measurements, calculated according to the formula:

$$\bar{C}_{mol} = \frac{1}{n} \sum_{i=1}^{n} C_{mol,i}. \tag{4}$$

The standard type B uncertainty was evaluated taking into account standard uncertainties caused by measurement instruments, reference data [9] and [10], as well as by various chemical factors (effects of oxygen, diffusion of the $Cr_2O_7^{2-}$ ions into the auxiliary cell chamber, presence of electroactive impurities in the electrolyte) using the following formula:

$$u_B = \sqrt{\sum_{j=1}^{l} \left(\frac{\partial f}{\partial x_j}\right)^2 \cdot u^2(x_j)} = \sqrt{\sum_{j=1}^{l} c^2(x_j) \cdot u^2(x_j)}. \tag{5}$$

where

x_j is the jth input value contributing to the standard type B uncertainty;
$u(x_j)$ is the standard uncertainty induced by the evaluation of the jth input value;
$c(x_j)$ is the sensitivity coefficient describing changes in output values under changes in the jth input value. Sensitivity coefficients $c(x_j)$ for each input value x_j contributing to u_B were determined as partial derivatives of the general measurement equation.

6 Evaluation of the CRM Instability

The ISO Guide 35 [15] distinguishes two main characteristics of CRM instability, namely short-term and long-term instability. In terms of CRMs for the composition of pure chemical substances, these characteristics are determined by various factors, including chemical and physical (adsorption e.g.) processes in solutions, package integrity, permeability characteristics of package material, exposure to such potentially harmful factors as temperature, UV light, etc. The assessment of the individual contribution of these factors is a hard task; therefore, the stability of CRMs is always assessed empirically for every CRM and package type. Thus, the standard uncertainty due to instability is evaluated by the standard deviation of the coefficient of the linear dependence of measurement results on time. This value is taken into account when the CRM shelf life is assigned. The long-term stability of the CRM for the composition of potassium dichromate solutions was evaluated by the method of accelerated aging according to the algorithm given in paragraph 5.3 of the document R 50.2.058-2007 "Recommendations for Metrology. State system for ensuring the uniformity of measurements. Estimation of uncertainties of CRM certified values" [16].

The aging temperature was taken to be (60 ± 2) °C, since the limit melting temperature of the insulating P-2 type paraffin equals 65 °C according to GOST 23683-89 "Solid petroleum paraffins. Technical conditions" [17].

The evaluation of uncertainties induced by long-term instability for the CRM under study is presented in Table 2.

Following all the measurements, the standard deviation of repeatability Sr was evaluated using the formula:

$$s_r = \sqrt{\frac{1}{n}\sum_{i=1}^{n}\frac{d^2}{2}} = 0.0003731 \quad (6)$$

The following linear model of the certified value D change at a time point t was suggested:

$$D = at, \quad (7)$$

Table 2 Evaluation of uncertainty induced by long-term instability for the CRM under study

Number of measurement, n	Measurement time, days	Measurement result, x_{0i}	Measurement result, x_i	Deviation d_i	$di \cdot t_i$
1	0	0.100489	0.100477	−0.000011	0
2	6	0.100532	0.100489	−0.000043	−0.000246
3	11	0.100903	0.100488	−0.000415	−0.00473
4	17	0.101203	0.100538	−0.000665	−0.011385
5	23	0.101814	0.100545	−0.001269	−0.028947

where a is the rate of the certified value change under conditions of intermediate precision.

The change of the certified value D_i at a time point t_i evaluated by measurement results was presented as

$$D_i = at_i + \varepsilon_i \tag{8}$$

where ε_i is the deviation of the ith measurement result under conditions of intermediate precision.

The coefficient a in Eq. (8) was determined according to the obtained value D_i by the least square method according to the formula

$$a = \frac{\sum_{i=1}^{n} D_i t_i}{\sum_{i=1}^{n} t_i^2} = -0.0000464 \tag{9}$$

The standard deviation for coefficient a, s_a was determined according to the formula

$$s_a = \sqrt{\frac{s_r^2}{n}} = 0.00001194, \tag{10}$$

The standard uncertainty induced by instability, u_{stab}, at a time point t (1 year) was evaluated by the formula

$$u_{stab} = s_a t = 4.4 \cdot 10^{-6} (0.0044\%) \tag{11}$$

Thus, the standard uncertainty induced by long-term instability for the period of 1 year was 0.0044%.

The short-term stability of the CRM for the composition of potassium dichromate solutions was assessed during the study of its long-term stability. The standard uncertainty due to the short-term CRM instability was taken to be numerically equal to that induced by the long-term CRM instability, since the critical (limit) conditions permissible during sample transportation corresponded to the maximum package integrity temperature of 60 °C.

7 Uncertainty Budget

An uncertainty budget for the establishment of the certified values of the CRM for the composition of potassium dichromate solution is given in Table 3.

The limits of relative error were evaluated using an algorithm given in GOST 8.736-2011 "State system for ensuring the uniformity of measurements. Multiple direct measurements. Methods for processing measurement results. Main principles" [18] and taking into account the characteristics of the CRM short-term and long-term stabilities.

Metrological characteristics of the developed CRM are presented in Table 4.

Both the relative expanded uncertainty ($k = 2$) and the limits of the relative error of the certified value ($P = 0.95$) did not exceed the value of 0.05%, thus satisfying the requirements of technical specifications formulated for the CRM development.

8 Conclusion

Following a series of experiments, a CRM for the composition of potassium dichromate solutions (0.1M $K_2Cr_2O_7$ CRM, UNIIM) was developed and investigated. This CRM was introduced into the State Register of Certified Reference Materials under the number GSO 10992-2017.[8] The developed CRM can be used for: transferring the mass (molar) concentration unit of a component to RMs and chemical reagents via oxidation-reduction reactions; the verification and calibration of MI, as well as control of metrological characteristics during MI trials including type approval; MI graduation; validation of measurement procedures; control of measurement result accuracy when determining the mass (molar) fraction and mass (molar) concentration of components in liquid and solid substances and materials.

The CRM is supplied in plastic bottles with screw caps, which are additionally sealed with paraffin or are equipped with shrink tubes. A 250 cm^3 sample of the CRM is presented in Fig. 4.

The developed CRM for the composition of potassium dichromate solution is a convenient instrument that can be used by testing laboratories both for controlling preparation of titrants for redox titration and for measuring target components in analyzed substances, thus significantly reducing the measurement preparation stage.

---

[8]GSO 10992-2017 Certified reference material for the composition of potassium dichromate solution (0.1M $K_2Cr_2O_7$, CRM UNIIM). Available at: http://www.fundmetrology.ru/09_st_obr/view.aspx?regn=%D0%93%D0%A1%D0%9E%2010992-2017.

Table 3 Uncertainty budget for the establishment of the certified value of the CRM for the composition of potassium dichromate solution

Type	Characteristics	Value		Uncertainty		Distribution	Sensitivity coefficient		Contribution to u_B	
		x_i	unit	u_i	unit		c_i	unit	$c_i u_i$	unit
A	**Repeatability**	0.100509	%	0.000008	mol/dm^3	N	1	1	8.53×10^{-6}	mol/dm^3
B	U_1	0.008450	V	7.47×10^{-5}	B	R	4.61×10^{-6}	(mol/dm^3)/V	3.44×10^{-10}	mol/dm^3
B	t'_1	202.8112	s	1.12×10^{-4}	s	R	5.65×10^{-10}	(mol/dm^3)/s	6.32×10^{-14}	mol/dm^3
B	t_1	133.8963	s	6.61×10^{-5}	s	R	-5.65×10^{-10}	(mol/dm^3)/s	3.74×10^{-14}	mol/dm^3
B	U_2	1.600249	V	5.85×10^{-4}	V	R	6.27×10^{-5}	(mol/dm^3)/V	3.67×10^{-8}	mol/dm^3
B	t_2	937.9749	s	9.00×10^{-5}	s	R	1.07×10^{-7}	(mol/dm^3)/s	9.63×10^{-12}	mol/dm^3
B	U_3	0.007947	V	7.28×10^{-5}	V	R	7.82×10^{-6}	(mol/dm^3)/V	5.70×10^{-10}	mol/dm^3
B	t_3	116.9748	s	7.07×10^{-5}	s	R	5.32×10^{-10}	(mol/dm^3)/s	3.76×10^{-14}	mol/dm^3
B	m	2.629851	g	1.44×10^{-4}	g	R	3.82×10^{-2}	(mol/dm^3)/g	5.51×10^{-6}	mol/dm^3
B	R	10.001409	Ohm	5.00×10^{-6}	Ohm	R	-1.00×10^{-2}	(mol/dm^3)/Ohm	5.02×10^{-8}	mol/dm^3
B	ρ	1.018394	g/cm^3	8.53×10^{-6}	g/cm^3	R	9.87×10^{-2}	(mol/dm^3)/(g/cm^3)	8.42×10^{-7}	mol/dm^3

(continued)

Table 3 (continued)

Type	Characteristics	Value		Uncertainty		Distribution	Sensitivity coefficient		Contribution to u_B	
		x_i	unit	u_i	unit		c_i	unit	$c_i u_i$	unit
B	F	96485.33289	C/mol	5.90×10^{-4}	C/mol	N	-1.04×10^{-6}	(mol/dm^3)/ (C/ mol)	6.15×10^{-10}	mol/dm^3
B	Diffusion	-	-	1.12×10^{-7}	mol/dm^3	R	1	-	1.12×10^{-7}	mol/dm^3
B	Purity of Ar (O_2)	-	-	6.01×10^{-12}	mol/dm^3	R	1	-	6.01×10^{-12}	mol/dm^3
B	Impurities	-	-	1.64×10^{-7}	mol/dm^3	R	1	-	1.64×10^{-7}	mol/dm^3
B	Influence of O_2	-	-	7.04×10^{-7}	mol/dm^3	R	1	-	7.04×10^{-7}	mol/dm^3
Certified value, C_{mol}									**0.100509**	mol/dm^3
Relative standard uncertainty type A, u_{A0}, %									0.0085	
Relative standard uncertainty type B, u_{B0}, %									0.0056	
Relative standard uncertainty induced by characterization, u_{char}, %									0.0102	
Relative standard uncertainty induced by long-term CRM instability, u_{stab}, %									0.0044	
Relative standard uncertainty induced by short-term CRM instability, $u_{s.stab}$, %									0.0044	
Relative combined standard uncertainty, u_{c0}, %									0.0119	
Relative expanded uncertainty, **U_0 ($k = 2$), %**									**0.024**	
U_1 is voltage during preliminary titration, V;										
U_2 is voltage during the main titration, V;										
U_3 is voltage during final titration, V;										
t'_1 is the total time of preliminary titration, s;										
t_1 is the time corresponding to the end point of preliminary titration, s;										
t_2 is the time of the main titration, s;										
t_3 is the time corresponding to the end point of the final titration, s.										

Table 4 Metrological characteristics of the CRM for the composition of potassium dichromate solution

CRM certified characteristic	CRM certified value	Relative expanded uncertainty of the certified value (k = 2), %	Limits of the relative error of the CRM certified value (P = 0.95), %
Molar concentration of potassium dichromate, mol/dm^3	0.100509	0.024	±0.026

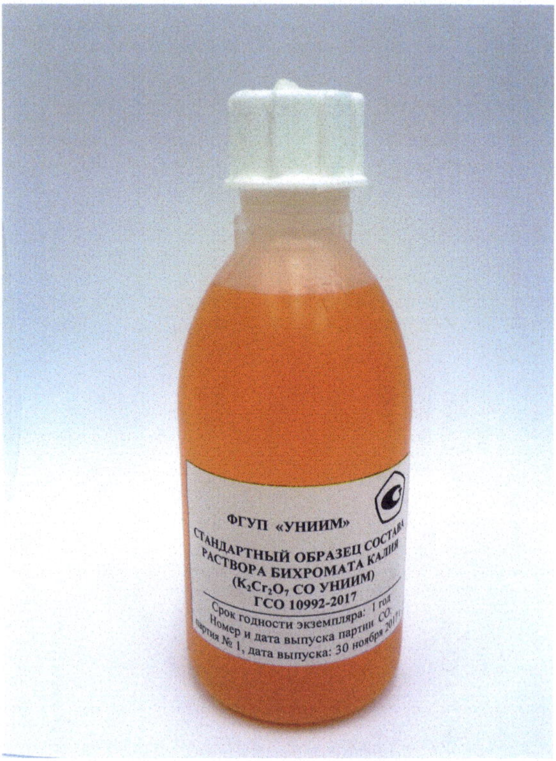

Fig. 4 A sample of the CRM for the composition of 0.1M potassium dichromate solution

Author Contribution Statement A. V. Sobina: development of the research concept, organization of the research, methodological support, preparation of the manuscript.

A. Yu. Shimolin: development of the research concept, performance of measurements and calculations, preparation of documents for the CRM certification, preparation of the manuscript.

References

1. Milton MJT, Quinn TJ (2001) Primary methods for the measurement of amount of substance. Metrologia 8:289–296. https://doi.org/10.1088/0026-1394/38/4/1
2. Mariassy M, Pratt KW, Spitzer P (2009) Major applications of electrochemical techniques at national metrology institutes. Metrologia 46:199–213. https://doi.org/10.1088/0026-1394/46/3/007
3. Terentiev GI, Sobina AV, Shimolin AJ, Zyskin VM (2014) Application of coulometric titration for the certification of primary reference materials of pure substances. AJAC 5: 559–565. https://doi.org/10.4236/ajac.2014.59063
4. Borges PP, Silva Junior WB, Gonzaga FB, Sobina AV, Shimolin AYu, Terentiev GI (2012) Coulometric determination of amount content of potassium dichromate for comparability assessment through a bilateral comparison. In: Proceedings of the XX IMEKO world congress metrology for green growth, Busan, 9–14 Sep 2012 (in Korea)
5. Máriássy M et al (2013) Key comparison CCQM-K96 determination of amount content of dichromate. Draft B report. Bratislava, Feb 2013
6. Terentiev GI, Gerasimova NL, Mamonova IS, Borozdina NF, Zyskin VM, Marina NV (2005) Transfer of standard titres to the category of certified reference materials is a new step in metrological assurance of analytical measurements. Ref Mater 1(2):17–22 (in Russ.)
7. Sobina AV (2014) Development of state primary standard units of mass fraction and mass concentration of the component in the liquid and solid substances and materials based on coulometric titration. Dissertation, D.I. Mendeleyev Institute for Metrology (VNIIM). Available at: http://vniim.ru/files/dis-sobina.pdf
8. Marinenko G, Taylor JK (1963) Precise coulometric titrations of potassium dichromate. J Res Nat Bur Stand A Phys Chem 67A(5):453–459
9. Atomic weights of the elements 2013 (IUPAC Technical Report) (2016) Pure Appl Chem 88(3):265–291. https://doi.org/10.1515/pac-2015-0305
10. Mohr PJ, Newell DB, Taylor BN (2016) CODATA recommended values of the fundamental physical constants: 2014. J Phys Chem Ref Data 45:043102. https://doi.org/10.1063/1.4954402
11. Shimolin AJ, Sobina AV, Zyskin VM (2017) Potassium iodate purity determination by high precision coulometric titration: new measurement procedure implementation. Chelyabinsk, 16–19 Oct 2017. IEEE, 17410812. https://doi.org/10.1109/uralcon.2017.8120729
12. Perelman VI (1963) Chemist quick reference. Scientific and technical publishing house of chemical literature, Moscow. (in Russ.)
13. GOST 4220-75 Reagents. Potassium bichromate. Specifications. Standartinform Publ., 2010. 6 p. (in Russ.)
14. JCGM 100:2008 Evaluation of measurement data—guide to the expression of uncertainty in measurement. JCGM, 2008. Available at http://www.bipm.org/utils/common/documents/jcgm/JCGM_100_2008_E.pdf
15. ISO Guide 35:2017 Reference materials—Guidance for characterization and assessment of homogeneity and stability. Available at https://www.iso.org/standard/60281.html
16. P.50.2.058-2007 State System for Ensuring Uniform Measurement. Evaluation of certified values uncertainties of reference materials. Standartinform Publ., 2008. 28 p. (in Russ.)
17. GOST 23683-89 Petroleum paraffin waxes. Specifications. Standartinform Publ., 2008. 14 p. (in Russ.)
18. GOST 8.736-2011 State system for ensuring the uniformity of measurements. Multiple direct measurements. Methods of measurement results processing. Main principles. Standartinform Publ., 2008. 24 p. (in Russ.)

Development and Testing of a Certified Reference Material of the Lead Mass Fraction in Solutions for Measurements Using Inductively Coupled Plasma Methods

Alexey A. Stakheev and Tatiana P. Stolboushkina

Abstract For the metrological assurance of the high-precision inductively coupled plasma (ICP) mass spectrometry and optical emission spectroscopy methods, the All-Russian Scientific Research Institute of Physical Technical and Radio Technical Measurements (VNIIFTRI) has launched works on the development of reference materials (RMs) of the mass fraction of metals in solutions specifically intended for measurements using inductively coupled plasma methods (ICP-CRM). The ICP-CRM under development was represented by a solution with the certified value of the lead mass fraction. The ICP-CRM was packed in high-density polyethylene bottles with the capacity of 4, 8, 15, 30, 60 and 125 cm^3. The certified value of the lead mass fraction in the solution was established by the preparation method and confirmed by the GET 217-2018 State Primary Standard for units of the mass fraction and mass (molar) concentration of inorganic components in aqueous solutions based on gravimetric and spectral methods. The permissible certified values of the lead mass fraction in the developed ICP-CRM are shown to range from 800 mg/kg to 1200 mg/kg. The uncertainty in the certified value is expressed as expanded uncertainty, $U = 0.5\%$, at 95% confidence and is calculated in accordance with ISO/IEC17025 according to GUM (Guidelines to Uncertainty in Measurement). The developed ICP-CRM solution of the lead mass fraction can be used for ensuring the metrological traceability of measurements in inorganic analyses using ICP-MS and ICP-OES to the GET 217-2018.

Keywords Inductively coupled plasma mass spectrometry · Inductively coupled plasma optical emission spectroscopy · Water · Aqueous solution · Standard · Reference material · Lead · Water analysis · Inorganic component

---

A. A. Stakheev · T. P. Stolboushkina (✉)
All-Russian Scientific Research Institute of Physical Technical and Radio Technical Measurements (VNIIFTRI), Moscow Region, Mendeleevo, Russia
e-mail: stolboushkina@vniiftri.ru

S. Medvedevskikh et al. (eds.), *Reference Materials in Measurement and Technology*,
https://doi.org/10.1007/978-3-030-32534-3_12

Abbreviations

GET 217-2018	the State Primary Standard for units of the mass fraction and mass (molar) concentration of inorganic components in aqueous solutions based on gravimetric and spectral methods
ICP CRM	Reference materials intended for use in measurements using inductively coupled plasma
ICP-MS	Inductively coupled plasma mass spectrometry
ICP-OES	Inductively coupled plasma optical emission spectrometry

1 Introduction

The monitoring of the content of inorganic components in various substances and materials is an urgent problem both in science and in industry. Under the contemporary conditions of technical progress and population growth, the issue of improving the quality control of various substances and materials is becoming increasingly important. Extensive branches of analytical chemistry are devoted to the measurement of a wide variety of components—from essential macro- and micro elements to highly toxic elements in diverse matrices. Such measurements are indispensable in the areas of environmental protection, health care, industrial safety and all those addressing the problem of human health. For example, the monitoring of water quality is known to have a positive effect on human health, thereby reducing health care costs.

Increasing requirements to the quality control of industrial production lead a higher demand for accurate measurement methods. When choosing an analytical method, chemists are guided by many parameters; however, the main selection criterion is for the method to provide the necessary range of measured concentrations and the required accuracy of measurement results. Along with that, analytical chemists consider whether the selected method is quick and convenient. Thus, the methods of inductively coupled plasma mass spectrometry (ICP-MS) and inductively coupled plasma optical spectroscopy (ICP-OES) are widely used because of such their advantages as the capability of multi-element analysis, low detection limits, a large linear concentration range and a high degree of automation when performing inorganic analyses. The efficiency of high-precision ICP-MS and ICP-OES is determined, to a large extent, by the quality and availability of certified reference materials (CRMs).

The aim of this work is the development and testing of a CRM for the mass fraction of lead in solutions for ensuring the metrological traceability of measurements performed using the ICP-MS and ICP-OES methods.

2 Literature Review

In recent years, the concept of metrological traceability in chemical measurements has become extremely important [1]. In accordance with [2], all chemical measurements in the Russian Federation should apply the standards of units traceable to the State primary standards of the corresponding units. Moreover, all measurements performed in the field of state regulation should also be traceable to State primary standards. In order to ensure the uniformity of measurements in the field of physico-chemical measurements of the composition of liquid and solid substances and materials, a set of State primary standards has been developed, which includes the State primary standard for the units of mass fraction and mass (molar) concentration of inorganic components in aqueous solutions based on gravimetric and spectral methods GET 217-2018.

The most rational way to achieve comparable and reliable measurement results by ICP-MS and ICP-OES in testing and analytical laboratories consists in establishing the metrological traceability of measurements using CRMs. Similar to other methods, ICP-MS and ICP-OES have their limitations [3–8], with the most problematic one being interference. Therefore, specific requirements in terms of impurity composition are imposed on CRMs intended for the calibration of ICP-MS and ICP-OES. For example, it is highly undesirable to use CRMs that either have been produced using glassware or are delivered in glassware. This material can be the reason of impurities, which significantly affect the accuracy of measurement results. The indication of the CRM impurity composition in their passports allows all possible sources of interference to be taken into account when measuring the target component.

Both ICP-OES and ICP-MS are broadly recognized as the methods permitting the achievement of highly accurate measurement results in elemental analysis [9–13]. Their low detection limits ensure the maintenance of the content of various elements at ppb and ppt levels. Chemical compounds that are particularly dangerous for the human health and known as cumulative toxicants include those of lead, cadmium and mercury [14]. These metals, even in trace amounts, can accumulate in the human body and cause serious health disorders. This is why lead was the first element, for which an ICP-CRM was developed at the VNIIFTRI.

3 Materials and Methods

The material chosen for the preparation of an ICP-CRM of the lead mass fraction was aqueous lead solution acidified with nitric acid. During the course of preliminary experiments, lead nitrate was chosen as a starting material. For storing, bottles made of high-density polyethylene with a capacity of 30, 60 and 125 cm^3 were used. A low-concentration aqueous solution of nitric acid was used as a solvent, which was believed to ensure the complete dissolution of the starting material and the stability of the developed ICP-CRM. Since the main sources of impurities in CRMs are those contained in water and reagents, a particular attention was paid to the process of their purification. To this end, lead nitrate, water and nitric acid were purified by a specifically developed method, a multi-stage purification system and a two-stage sub-boiling distillation method, respectively. In addition, a particular attention was paid to the selection and pre-cleaning of polymer laboratory ware.

In order to assess the purity of the starting material, the "100% minus the amount of impurities" method was chosen. The mass fraction of the main component (W) was calculated by Eq. (1):

$$W = 100\% - \sum_i \omega_i - \sum_j \frac{LOD_j}{2}, \tag{1}$$

where

ω is the mass fraction of detected impurities, %;
LOD is the limit of detection of undetected impurities, %

The impurity composition of the solvents (nitric acid and water) was assessed after their purification using the GET 217-2018. A scheme describing the CRM preparation from lead nitrate is presented in Fig. 1.

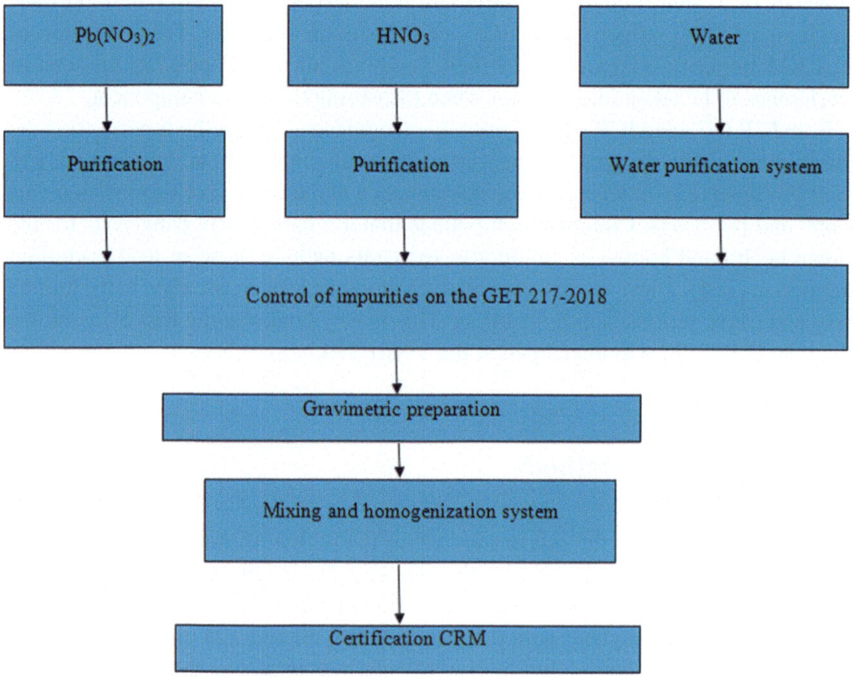

Fig. 1 Block diagram describing the preparation of a CRM of the lead mass fraction

The mass fraction of the certified component (lead) in the solution was estimated by Eq. (2):

$$C_{am} = \frac{W \cdot c_c \cdot m_c + c_s \cdot (m_{solution} - m_c)}{m_{solution}}, \quad (2)$$

where

W is the mass fraction of the main component in the starting material;

c_c, c_s are the mass fractions of the certified component in the starting material and solvent, respectively;

m_c, m_p, $m_{solution}$ are the weights of the starting material, solvent and RM samples.

4 Results and Discussion

The developed ICP-CRM for the lead mass fraction in aqueous solutions is packed in high-density polyethylene bottles with a capacity of 30, 60 and 125 cm^3 (Fig. 2).

The reliability of the certified value of the lead mass fraction in the developed ICP-CRM was established by two independent methods, including the analytical experimental method of gravimetric preparation and the method of measuring the lead mass fraction using the capacity of the GET 217-2018. The measurement results obtained using both abovementioned methods are shown to be in good agreement within the limits of their uncertainties, as well as with the results of comparative measurements carried out for the developed ICP-RM and the RMs of other manufacturers. The certified value of the developed CRM is taken to be that obtained by the analytical experimental method of gravimetric preparation.

The measurement uncertainty of the certified value of the analyte mass fraction in solutions was calculated in accordance with [15, 16]. The uncertainty components for

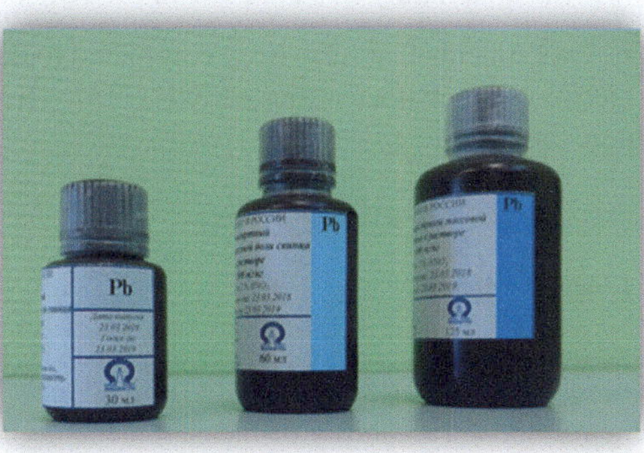

Fig. 2 ICP-CRM of the lead mass fraction

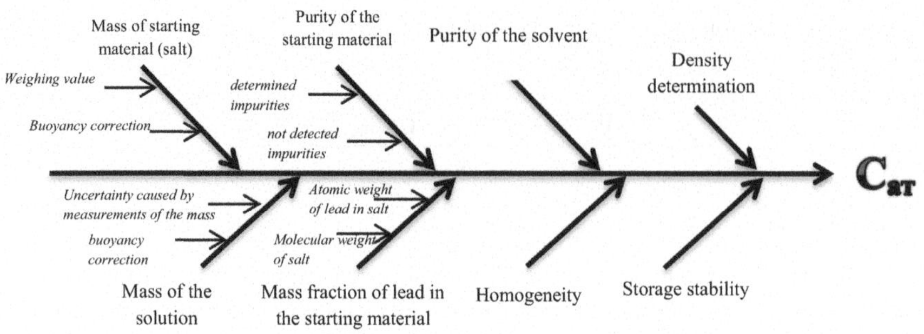

Fig. 3 A cause-and-effect diagram for the components of measurement uncertainty associated with gravimetric preparation

Table 1 The metrological characteristics of the ICP-CRM lead mass fraction

Certified characteristic	The shelf life	Interval of certified values	Expanded uncertainty, U_0 ($k = 2$; $P = 0.95$)
Lead mass fraction	12 months	(800–1200) mg/kg	0.5%

the gravimetric preparation of CRM solutions for measuring the lead mass fraction are presented as a cause-and-effect diagram in Fig. 3.

The total standard uncertainty of the certified value of the mass fraction of a component in the ICP-CO is determined by Eq. (3):

$$u_c = \sqrt{u_{char}^2 + u_{hom}^2 + u_{stab}^2}, \qquad (3)$$

where

u_{char} is the uncertainty associated with the method of establishing the certified value of the lead mass fraction in the ICP-CRM,
u_{hom} is the uncertainty associated with the ICP-CRM heterogeneity,
u_{stab} is the uncertainty associated with the ICP-CRM instability.

The experiments that have been completed to date have established the shelf life and metrological characteristics of ICP-CRM. The characteristics of ICP-CRM are given in Table 1.

5 Conclusion

The developed ICP-CRM solution of the lead mass fraction can be used for ensuring the metrological traceability of measurements in inorganic analyses using ICP-MS and ICP-OES. The certified value expressed in the mg/kg units allows a more accurate, gravimetric method of solution preparation to be used, whereas the assigned solution density allows, if necessary, a more familiar and fast volumetric method to be used instead.

Author Contribution Statement Stakheev A. A.: development of the research concept and the article concept, critical revision of the manuscript.
Stolboushkina T. P.: conducting experiments, obtaining experimental data, development of preparation and measurement methods, literature review.

References

1. De Bièvre P, Dybkaer R, Fajgelj A, Hibbert DB (2011) Metrological traceability of measurement results in chemistry: Concepts and implementation (IUPAC technical report). Pure Appl Chem 83(10):1873–1935. https://doi.org/10.1351/PAC-REP-07-09-39
2. Federal law On ensuring the uniformity of measurements. No FZ-102 of 26.06.2008. http://fundmetrology.ru/depository/01_npa/102-fz_2015.pdf (in Russ.)
3. Turner BL, Bielnicka AW, Dalling JW, Wolf JA (2013) Interference by iron in the determination of boron by ICP-OES in mehlich-III extracts and total element digests of tropical forest soils. Commun Soil Sci Plant Anal 47(21):2378–2386
4. Mozafar A, Oertli JJ, Tuor D, Roesch T (1990) Calibration of ICP to reduce the interference of Na with the determination of K. Commun Soil Sci Plant Anal 21(13):1607–1617
5. Zbíral J (1996) Determination of potentially dangerous elements in soil extracts by atomic emission spectrophotometry-ICP: Correction of interferences. Commun Soil Sci Plant Anal 27(5):1527–1536
6. Morgalla MH, Mettenleiter H, Katzenberger T (1999) ICP measurement accuracy: the effect of temperature drift. Design of a laboratory test for assessment of ICP transducers. J Med Eng Technol 23(1):10–14. https://doi.org/10.1080/030919099294375
7. Zhong J, Dujovny M, Park HK, Perez E, Perlin AR, Diaz FG (2003) Advances in ICP monitoring techniques. Neurol Res 25(4):339–350
8. Pupyshev A, Surikov VT (2006) Inductively coupled plasma–Mass spectrometry. Formation of ions. UrO RAN, Ekaterinburg
9. Chung YS, Chung YJ, Cho KH, Lee JH Determination of trace and toxic elements in Korean rice CRM by INAA, ICP and AAS. J Radioanal Nucl Chem 215(1):129–134. https://doi.org/10.1007/bf02109888
10. Fernández-Turiel JL, Llorens JF, Roig A, Carnicero M, Valero F (2000) Monitoring of drinking water treatment plants using ICP-MS. Toxicol Environ Chem 74(1–2):87–103. https://doi.org/10.1080/02772240009358871
11. Šťastná M, Němcová I, Zýka J (1999) ICP-MS for the determination of trace elements in clinical samples. Anal Lett 32(13):2531–2543
12. Masson P, Dalix T, Bussière S (2010) Determination of major and trace elements in plant samples by inductively coupled plasma–mass spectrometry. Commun Soil Sci Plant Anal 41(3):231–243. https://doi.org/10.1080/00103620903460757
13. Konda SK, Kiran B, Hardaway CJ, Sneddon J (2014) Determination of copper, iron, lead, magnesium, manganese, and potassium in rice from India and United States by inductively coupled plasma-optical emission spectrometry. Instrum Sci Technol 42(6):646–651. https://doi.org/10.1080/10739149.2014.919510
14. Stepin BD, Tsvetkov AA (1994) Inorganic chemistry. Higher school, Moscow
15. Ellison R, Williams A (eds) (2012) Quantifying uncertainty in analytical measurement. In: Eurachem/CITAC guide, 3rd edn. https://www.eurachem.org/index.php/publications/guides/quam
16. GOST R 54500.3-2011. Uncertainty of measurement. Part 3. Guide to the expression of uncertainty in measurement. Standardinform, Moscow, 2012 (in Russ.)

Radiation-Sensitive Film Compositions for Measuring Absorbed Doses Within the 100–1000 Gy Range

Vladimir P. Tenishev and Ivan A. Emelyanenko

Abstract This work was aimed at investigating the spectral features and dose sensitivities of film dosimetry systems developed at the All-Russian Scientific Research Institute of Physical Technical and Radio Technical Measurements (VNIIFTRI) and approved as absorbed-dose reference materials. The properties of new radiation-sensitive compositions for high-intensity ionising radiation in the absorbed-dose range from 100 to 1000 Gy were studied. Dosimetric measurements in this range are in demand in the sphere of ionising radiation technologies used in agriculture and food-processing industries.

Keywords Radiation processing technologies · Radiation processing unit (RPU) · Ionising radiation · Absorbed dose (AD) · Dosimetry · Radiation-sensitive compositions · Post-effect · Heat treatment

1 Introduction

When processing products with ionising radiation, technological goals can only be achieved provided that the following parameters of a radiation processing unit (RPU) have been correctly set: radiation energy, conveyor speed, orientation of the product with respect to the irradiator, etc. This problem can be solved by appropriate dosimetry [1–8] with the absorbed dose mapping across the entire volume of the irradiated product. Thus, the corresponding parameters of the RPU are set on the basis of the obtained dosimetric data. The absorbed-dose range in the analysed product must meet the regulatory requirements for absorbed doses defined in the technological regulations for the irradiation of a particular product in accordance with the requirements for ensuring the quality and safety of consumer products [9].

Certified reference materials (CRMs) on the basis of coloured radiation-sensitive films (CRFs) developed by VNIIFTRI are currently used for the dosimetric control of RPUs applied in the sphere of radiation processing technologies in the Russian

V. P. Tenishev (✉) · I. A. Emelyanenko
All-Russian Scientific Research Institute of Physical Technical and Radio Technical Measurements (VNIIFTRI), Moscow Region, Mendeleevo, Russia
e-mail: tenishev@vniiftri.ru

S. Medvedevskikh et al. (eds.), *Reference Materials in Measurement and Technology*,
https://doi.org/10.1007/978-3-030-32534-3_13

Federation. These reference materials are traceable to the State Primary Special Measurement Standard (GET 209-2014) [10], which defines the unit of absorbed dose rate (ADR), allowing absorbed doses of high-intensity radiation to be measured in a range from 1 to 200 kGy along with the corresponding metrological characteristics.

However, the dose measurement range of absorbed-dose CRMs is currently insufficient for carrying out proper dosimetry in the course of irradiating food and agricultural products. Depending on the technological aim, absorbed-dose values may range from tens of Gy to several tens of kGy [6], which requires the development of new dosimetric systems characterized by increased sensitivity for measuring the absorbed dose of high-intensity ionising radiation in the range of 100–1000 Gy.

When irradiated in a transportation container, the non-uniformity of absorbed doses in certain types of products satisfies the condition $D_{max}/D_{min} \leq 3$. Under such conditions, the absorbed-dose distribution in the irradiated products established as a result of dose mapping must strictly satisfy the condition

$$D_{\min} \leq D_{irrad} \leq D_{\max},$$

where:

- D_{min} is the minimum value of the absorbed dose, at which the technological aim can still be achieved during product irradiation;
- D_{irrad} is the range of absorbed doses established as a result of dose mapping in the irradiated products;
- D_{max} is the highest value of the absorbed dose, at which a qualitative change occurs in the physical and technical properties of the product as a result of its exposition to ionising radiation.

Limitations imposed on the absorbed-dose value in the irradiated product require the use of dosimetric systems with established metrological characteristics, which are traceable to a national measurement standard across the entire permissible absorbed dose range [1–8]. Such systems permit the quality and safety of products to be preserved, along with the uniformity of absorbed dose measurements when processing products using ionising radiation in the Russian Federation [11, 12].

The experience of VNIIFTRI in the field of medical device sterilisation [9, 10] using electron accelerators and radioactive sources confirms the feasibility of dosimetric approaches using radiation-sensitive films. These approaches are characterised by convenience and low cost, as well as by the simplicity of measurements and a high spatially-resolved (up to 1 mm) mapping of the absorbed dose in the irradiated products. However, current film dosimetry systems based on radiation-sensitive compositions created by national institutes in the USA, Canada, the UK, the Netherlands, China and Russia are unable to provide radiation processing technologies with absorbed-dose measurements in the 100–1000 Gy range.

The creation of new film-based instruments measuring absorbed doses, which are sensitive in the specified absorbed dose range and stable to various input parameters, depends on the physical and chemical nature of radioactive transformations in multicomponent complex polymer compositions, dyes and sensitisers. This work does

Table 1 Radiation-sensitive compositions under study

Radiation-sensitive composition	Absorbed dose range, kGy	Radiation type	Abbreviation
Copolymer with 4-Dimethylaminoazobenzene	(1–10) kGy (200–1000) Gy	γ, e^-, β	CRF-2
Copolymer with Rhodamine C	(100–1000) Gy		

not consider the uncertainty budget of film dosimetry systems, as this would require a separate and detailed study into sources affecting the metrological characteristics of measuring instruments.

In the present work, we set out to assess the possibility of developing new film dosimetry systems for measuring absorbed doses of ionising radiation across the 100–1000 Gy range.

2 Materials and Methods

2.1 Radiation-Sensitive Compositions

Table 1 presents the content and abbreviations of radiation-sensitive compositions studied in the present work. CRFs were prepared from the listed compositions in accordance with a specific technology (TU 2379-026-13271746-2006). CRFs were obtained by applying a radiation-sensitive layer (that records absorbed doses) of the corresponding copolymer (Table 1) onto a transparent polyethylene terephthalate (PET) film base (~100 μm thick). The PET base was selected, since it had proven suitable for use as a radiation-sensitive layer in a number of widely used film dosimeters operating in the range of 1–200 kGy. The precise structure of the radiation-sensitive compositions is not specified here, since it is a proprietary right of the owner.

2.2 Experimental Equipment

The experimental film samples were irradiated using the following certified systems: MRH-γ-100 (radiation of radionuclides ^{60}Co with an energy of 1.25 MeV, dose rate of ~3 Gy/s) and LMB-γ-1M (radiation of radionuclides ^{137}Cs, with an energy of 0.65 MeV, dose rate of ~0.5 Gy/s). These systems are included in the GET 209-2014 State Primary Special Measurement Standard for the absorbed dose rate unit of intense photon, electron and beta radiation for radiation technologies [10].

The optical density (OD) of the studied film samples was measured prior and following irradiation using an Analytik Jena SPECORD 210 Plus spectrophotometer.

2.3 Sample Preparation and Measurement Procedure

Film sheets were prepared by applying a radiation-sensitive layer uniformly onto a PET base, with the thickness of the layer ranging from (10 ± 1) to (60 ± 3) μm. The film sheets were subsequently dried in a thermal oven at ~60 °C for 4–5 h. Afterwards, working samples (10 × 25 mm in size) were cut from these film sheets in accordance with the dimensions of spectrophotometer film holders. The OD of the non-irradiated films was monitored first visually and then using the spectrophotometer in order to select non-irradiated reference samples. For subsequent exposure to radiation doses, the stacks of 5–6 10 × 25 mm samples were placed into the certified area of the gamma-ray chamber. The OD of the films irradiated by various doses was measured using the spectrophotometer relative to the OD of non-irradiated reference samples.

3 Results

3.1 Spectral and Dosimetric Characteristics of the Radiation-Sensitive Composition of the Reference Material CRM AD(E)-1/10

The certified absorbed-dose reference material CRM AD (E)-1/10 (GSO 8916-2007) having a radiation-sensitive layer of (30 ± 3) μm is used as a chemical film dosimeter of ionising radiation in the absorbed-dose range of 1–10 kGy. At absorbed-dose values of less than 1 kGy, this dosimeter is characterised by low sensitivity and high measurement error. Therefore, for the purposes of this research, film samples with radiation-sensitive layers of different thickness were produced on the basis of the radiation-sensitive composition CRM AD (E)-1/10.

Figure 1 displays the dependence of the relative OD on the thickness of the radiation-sensitive layer of CRM AD(E)-1/10 under irradiation with a dose of ~1000 Gy. Measurements were performed for each thickness value of the radiation-sensitive layer (specified in Fig. 1) using five samples. The data presented in the figure shows the OD to increase linearly depending on the thickness of the radiation-sensitive layer in the region from ~34 to ~56 μm.

Figure 2 shows the measured spectral and dosimetric characteristics for the radiation-sensitive layer of CRM AD (E)-1/10 with a thickness of 56 ± 3 μm following irradiation with doses ranging from 200 to 1000 Gy. As can be seen from Fig. 2, the film sensitivity is sufficient for measuring the absorbed dose in the required

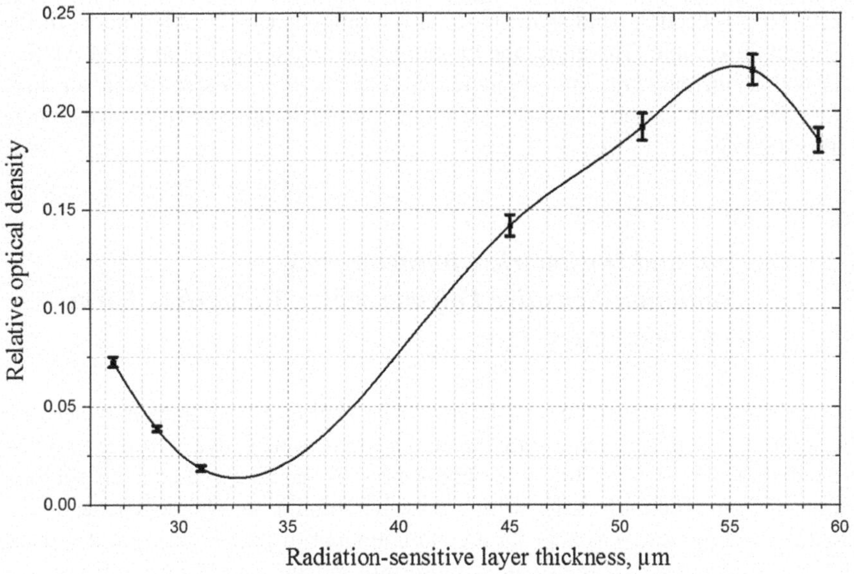

Fig. 1 Dependence of the CRF-2 sensitivity on the thickness of the radiation-sensitive layer at a wavelength of 550 ± 3 nm

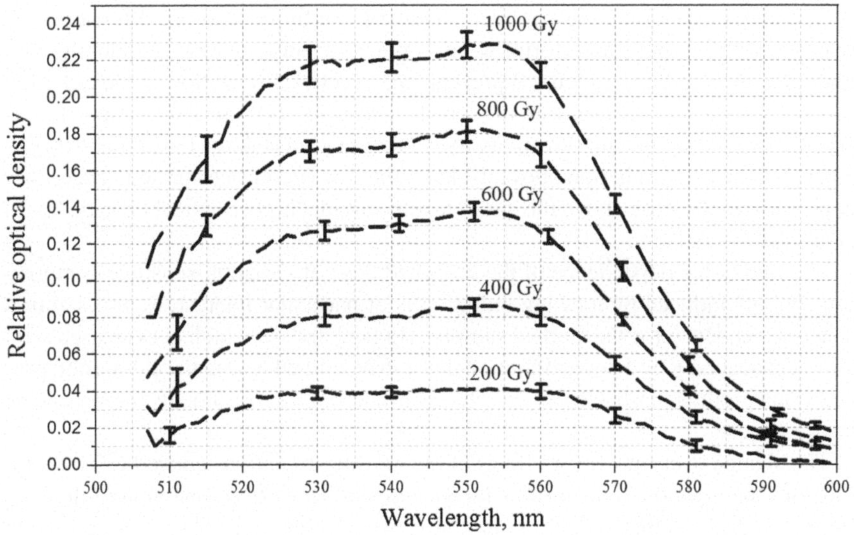

Fig. 2 Optical density of CRF-2 having a radiation-sensitive copolymer layer with an azo dye 4-Dimethylaminoazobenzene (~56 μm thick) after being irradiated with different doses

range. The absorption band shape in the given range is largely independent of the absorbed dose value. Therefore, the radiation-sensitive composition CRM AD(E)-1/10 with an increased thickness or increased concentration of the radiation-sensitive layer components can be recommended for measuring the absorbed dose in the range ~200–1000 Gy.

3.2 Spectral and Dosimetric Characteristics of a Radiation-Sensitive Polymer with a Colourless Form of the Rhodamine C Dye

An experimental composition was obtained from a halogen-containing polymer with the addition of a plasticiser, a stabiliser and a Rhodamine C dye sensitive to ionising radiation. The composition was obtained by mixing a neutral aqueous solution extracted from a colourless dye with a polymer solution to a homogeneous mass. The resulting mixture was kept for a certain amount of time in an open dry place and was subsequently poured onto a uniformly moving carrier tape substrate. In order to obtain samples of various thickness values, the speed of the moving carrier tape was changed. Prior to irradiation, the composition was colourless and optically transparent in the visible area of the spectrum. After irradiation with ^{60}Co or ^{137}Cs gamma-rays of the measurement standard, the film containing Rhodamine C assumed a bright and stable form, whose optical density increased gradually with increased absorbed dose. Figure 3 presents the spectral distribution of the radiation sensitivity for a composition with the sensitive layer of ~60 μm. This composition is seen to exhibit a significant sensitivity to irradiation even at doses lower than 100 Gy. When exposed to irradiation above 10 kGy, the dye becomes saturated. Therefore, an increase in the dose above this point results not in a higher optical density, but in its gradual degradation.

The functional dependence of the absorbed dose of radiation-sensitive compositions on OD induced by ionising radiation is of particular interest. Figure 4 shows such a dependence for the studied samples (Figs. 2 and 3), which is expressed with sufficient accuracy by a power function that can be refined during the calibration process when producing a batch of film reference materials as instruments for measuring absorbed dose.

The change over time (post-effect) of the ionising-radiation-induced OD of compositions having radiation-sensitive layers was studied by heat treatment of the irradiated samples. To this end, the radiation-sensitive films were placed in an oven at a certain temperature following irradiation with various doses. The measurement results for absorbed dose under heat treatment show that the post-effect is significantly reduced following heat treatment within the temperature range of 60–90 °C for up to 30 min. At the same time, the initial OD decreases as compared to the measured OD immediately following irradiation and prior to heat treatment, which

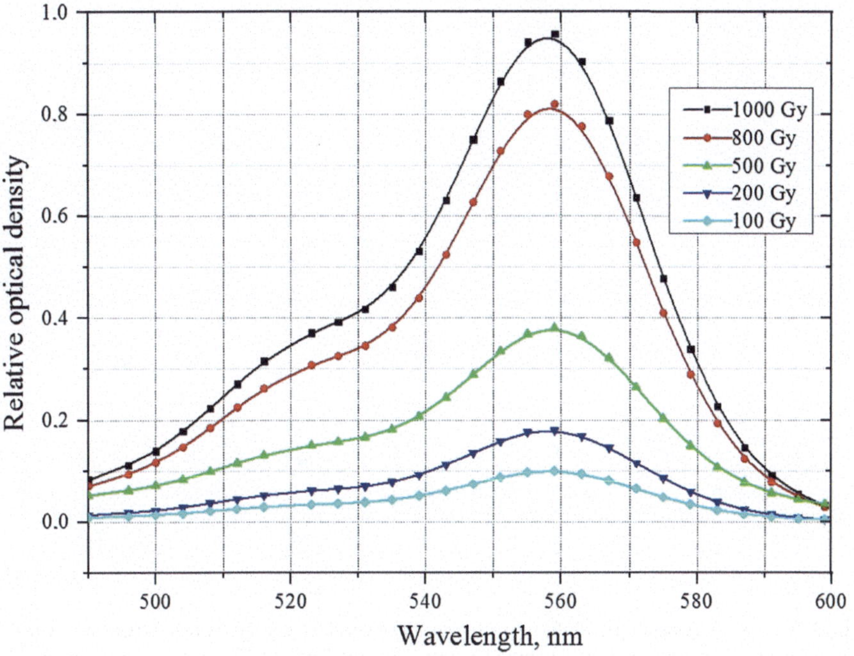

Fig. 3 Spectral dependence of optical absorption on the ionising-radiation dose for a radiation-sensitive composition (~60 μm thick) on the basis of Rhodamine C

should be taken into account when producing film instruments for measuring absorbed dose.

4 Conclusion

In the Russian Federation, the use of radiation processing technologies in agro-industrial and food industries is increasingly attracting research attention [11, 12]. According to [1–8], absorbed-dose measurements are necessary for controlling the operation of RPUs, which can ensure the quality and safety of irradiated products. For these purposes, instruments capable of measuring the absorbed dose of 100 Gy and above are in demand.

The radiation-sensitive compositions studied in this work permit the development of instruments for measuring the absorbed dose in the 100–1000 Gy range of high-intensity photon and electron radiation, which is widely used in radiation technologies.

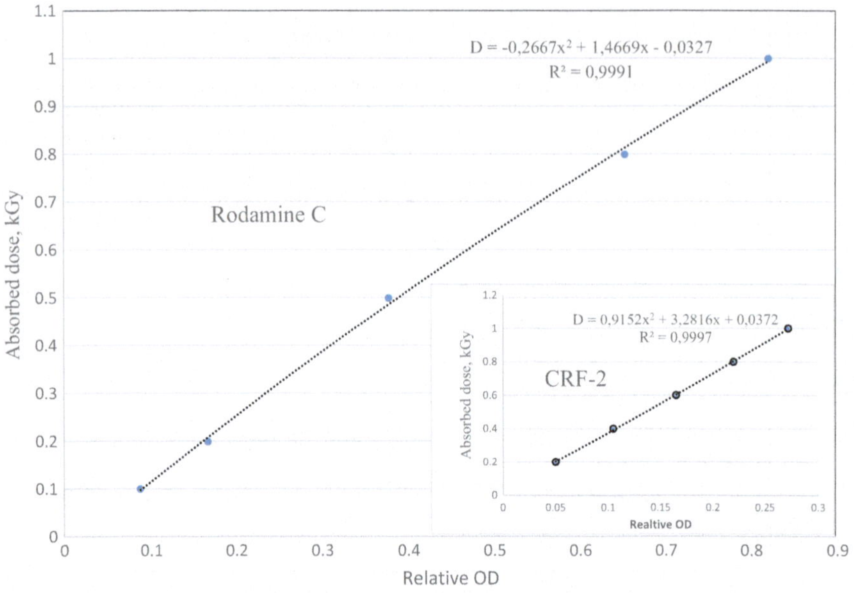

Fig. 4 Functional dependence of the absorbed dose D on optical density for two radiation-sensitive compositions on the basis of Rhodamine C (~60 μm) and CRF-2 (~56 μm) in accordance with figures 2–3

Acknowledgements The authors are grateful to the staff of the Laboratory of Technological Dosimetry, VNIIFTRI, for their assistance in experimental studies, processing of measurement results and constructive comments when preparing this work for publication.

Author Contribution Statement Tenishev V.P.—60%, Emelyanenko I.A.—40%.

References

1. ISO 11137-1:2006 Sterilization of health care products—Radiation—Part 1: Requirements for development, validation and routine control of a sterilization process for medical devices
2. GOST ISO 11137-1-2011 Sterilization of health care products—Radiation—Part 1: Requirements for development, validation and routine control of a sterilization process for medical devices. Standartinform, Moscow, 2013, 36 p (in Russ.)
3. ISO 11137-3:2017 Sterilization of health care products—Radiation—Part 3: Guidance on dosimetric aspects of development, validation and routine control of a sterilization process for medical devices
4. GOST R ISO 11137-3-2008 Sterilization of health care products—Radiation—Part 3: Guidance on dosimetric aspects. Standartinform, Moscow, 2009, 19 p (in Russ.)
5. ISO 14470:2011 Food irradiation—Requirements for the development, validation and routine control of the process of irradiation using ionizing radiation for the treatment of food
6. GOST ISO 14470-2014 Food irradiation—Requirements for the development, validation and routine control of the process of irradiation using ionizing radiation for the treatment of food. Standartinform, Moscow, 2015, 27 p (in Russ.)

7. ISO/ASTM 51900:2009 Guide for dosimetry in radiation research on food and agricultural products
8. GOST ISO/ASTM 51900 Guide for dosimetry in radiation research on food and agricultural products. Standartinform, Moscow, 2014, 37 p (in Russ.)
9. Generalova VV et al (2007) Ensuring the uniformity of measurements in radiation technologies. VNIIFTRI, Mendeleevo, 255 p. (in Russ.)
10. Aleykin VV, Generalova VV, Gromov AA, Gurskiy MN, Zhanzhora AP, Emelyanenko IA, Kovalenko OI, Tenishev VP (2015) State primary special measurement standard for the unit of absorbed dose rate of intense photon, electron and beta radiation for radiation technologies GET 209-2014. Almanac of modern metrology, VNIIFTRI, Mendeleevo. 5:54–74 (in Russ.)
11. Kozmin GV, Geraskin SA, Sanzharova NI (eds) (2005) Radiation technologies in agriculture and food industry. Obninsk, 400 p (in Russ.)
12. Radiation technologies in agriculture and food industry: Current state and prospects. Proceedings of the International Research and Practice Conference. Obninsk, 2018, 356 p. (in Russ.)

Developing a Reference Material Set for the Magnetic Properties of NdFeB Alloy-Based Hard Magnetic Materials

Ekaterina A. Volegova, Tatiana I. Maslova, Vladimir O. Vaskovskiy and Aleksey S. Volegov

Abstract This paper is devoted to the development of reference materials (RMs) for the magnetic properties of hard magnetic materials on the basis of $Nd_2Fe_{14}B$, a highly anisotropic intermetallic compound. Permanent magnets are widely used in different areas of technology, thus requiring measurements of major hysteresis loop for the correct calculation of magnetic system parameters. The main sources of error when measuring major hysteresis loops are given. The practical impossibility of verifying blocks of magnetic measuring systems element-by-element is noted. Nd–Fe-B permanent magnets were selected as the material for developing the RMs. RM certified values were established using a CYCLE-3 apparatus included in the GET 198-2017 State Primary Measurement Standard for units of magnetic loss power, magnetic flux density of constant magnetic field in a range from 0.1 to 2.5 T and magnetic flux in a range from 1×10^{-5} to 3×10^{-2} Wb. Based on the experimentally obtained major hysteresis loops, the magnetic characteristics were evaluated, the interval of permitted certified values was set, the measurement result uncertainty of certified values was estimated, the RM validity period was established and the first RM batch was released. On the basis of conducted studies, the RM type for magnetic properties of NdFeB alloy-based hard magnetic materials was approved (MS NdFeB set). The developed RM set was registered under the numbers GSO 11059-2018/GSO 11062-2018 in the State RM Register of the Russian Federation.

Keywords Type approved reference material · Remanence · Remanent magnetization · Coercivity · Hard magnetic materials · State primary measurement standard

E. A. Volegova (✉) · T. I. Maslova · A. S. Volegov
Ural Scientific Research Institute for Metrology (UNIIM), Yekaterinburg, Russia
e-mail: Ekaterina.Volegova@gmail.com

E. A. Volegova · V. O. Vaskovskiy · A. S. Volegov
Ural Federal University (UrFU), Yekaterinburg, Russia

S. Medvedevskikh et al. (eds.), *Reference Materials in Measurement and Technology*,
https://doi.org/10.1007/978-3-030-32534-3_14

1 Introduction

The level of technical and technological development in a given society is to a large extent determined by the functional and structural properties of the materials used. An important place among them is occupied by hard magnetic materials, whose products (permanent magnets) provide permanent magnetic fields that do not require an energy input. Hard magnetic materials are most widely used in the field of the mutual conversion of mechanical and electrical energy, as well as for the creation of magnetic fields having particular configurations. A significant increase in the production of brushless DC electric motors and electric generators having permanent magnet excitation both in Russia and abroad necessitates the control of the properties of permanent magnets that constitute such electric devices. The hysteretic magnetic properties of products from hard magnetic materials must be measured in a closed magnetic circuit using accurate magnetic measuring systems. Complete information on the major hysteresis loop of magnets allows an accurate calculation of the output parameters of magnetic systems and electrical devices at the stage of their development and, if necessary, a selective assembly of the magnetic system taking the batch magnet hysteresis property variation into account.

For the majority of permanent magnets, it is sufficient to indicate only the magnetic hysteresis loop point characteristics: remanence, B_r, T; remanent magnetization, M_r, kA/m; normal coercivity, H_{cB}, kA/m; intrinsic coercivity, H_{cM}, kA/m; the maximum energy product, $(BH)_{\text{max}}$, kJ/m^3.

As a rule, the properties of permanent magnets are measured in a closed magnetic circuit using compensated coils and an electromagnet as the magnetizer and circuit closer. Type approval testing, verification and calibration of such measuring instruments is rather difficult owing to the unique distribution of magnetic fields in every electromagnet at different electric current values and pole gap sizes, as well as differences in the geometric parameters of compensated coils from different producers. Currently, metrological support for measurements of the magnetic characteristics of permanent magnets is provided predominantly by means of element-by-element verification of magnetic measuring system blocks. In this case, it is necessary to consider a number of factors that influence the measurement results: magnetic field inhomogeneity in the working area; non-parallelism of the pole tip end surfaces; imperfections of the magnetic circuit, including scratches on the pole tip surface when carrying out measurements; magnetic viscosity of the material; the degree of coil compensation; the non-linearity of fluxmeters, etc. Identifying and taking account of such factors for each particular measuring system is rather labour-intensive and time-consuming. A more effective way to address this issue consists in developing and using type-approved reference materials (RMs) for ascertaining the magnetic properties of hard magnetic materials. The use of an approved type allows a reduction of the time input and economic expenditures allocated to testing for type approval, MI verification and calibration, as well as an increase in work efficiency by several times.

The purpose of this work is to carry out research in order to develop an RM set for the magnetic properties of Neodymium-Iron-Boron (NdFeB) alloy-based hard magnetic materials.

2 Materials and Measurement Methods

The RMs consist of Nd–Fe-B alloy-based cylinders containing heavy rare earth metals. The RMs were synthesized using sintering technology, including alloy melting, three-stage grinding, pressing the resulting alloy powder in an axial magnetic field, followed by sintering under vacuum or in an inert gas medium, and low-temperature annealing. The RMs have a nominal length of 10 mm and nominal diameters of 7-, 12-, 22- and 36-mm respectively. Due to the high brittleness of the magnets, the RMs were reinforced with an external austenitic stainless-steel ring of the same height as the hard-magnetic material in order to prevent their destruction under the action of compressive stresses caused by the electromagnet pole tips during the measurement process.

The measurement of RM magnetic properties and determination of the certified characteristics were performed in a closed magnetic circuit using the CYCLE-3 reference system included in the GET198-2017 State Primary Measurement Standard for the units of: magnetic loss power, magnetic flux density of constant magnetic field in a range from 0.1 to 2.5 T and magnetic flux in a range from 1×10^{-5} to 3×10^{-2} Wb, whose diagram and metrological characteristics are presented in [1]. The magnetic flux penetrating the RM was measured by means of an encircling coil. The value of the magnetic flux, which is coupled to all measuring coil turns, Ψ, Wb, is expressed as

$$\Psi = \Phi W = W \iint\limits_S \boldsymbol{B} d\boldsymbol{S}, \tag{1}$$

where W is the number of coil turns; dS is the area element, m^2; H is the magnetic field strength, A/m; $B = \mu_0 (H + M)$ is the magnetic flux density, T; M is the magnetization, A/m.

Magnetic field strength in an RM is determined in accordance with the electromagnetic field boundary conditions $H_{\tau 1} = H_{\tau 2}$ (the tangential component of the magnetic field strength is continuous when crossing the dividing boundary between two media in the absence of electric currents). In view of low air magnetic susceptibility $\chi \sim 10^{-6}$, $\mu_0 H = B$. Therefore, the formula for calculating the magnetization, M, A/m is as follows

$$M = 4\frac{\Psi - \mu_0 H K_{coil}}{\mu_0 \pi d_{RM}^2 W}, \tag{2}$$

where K_{coil} is the measuring coil constant, m^2; μ_0 is the magnetic constant, H/m; d_{RM} is the RM diameter, m.

The calculation of RM magnetic induction, B_{RM}, T is performed in accordance with

$$B_{RM} = \mu_0(H_{RM} + M_{RM}), \tag{3}$$

where B_{RM} is the RM magnetic flux density, T; H_{RM} is the magnetic field strength in the RM, A/m; M_{RM} is the reference material magnetization, A/m.

The magnetic characteristics of the RMs were obtained by analysing the magnetization-based RM major demagnetization curve *M* vs. *H* and RM major demagnetization curve *B* vs. *H*.

The remanent magnetization was determined on the basis of the magnetization-based RM major demagnetization curve *M* vs. *H* as the RM magnetization value following premagnetization to the state of technical saturation with the magnetic field decreasing in the RM to zero.

The remanence was determined on the basis of the major demagnetization curve *B* vs. *H* as the RM induction value following premagnetization to the state of technical saturation with the magnetic field decreasing in the RM to zero.

The intrinsic coercivity was determined on the basis of the magnetization-based RM demagnetization limit curve as the quantity equal to the magnetic field strength required for changing the magnetization from remanent magnetization to zero.

The normal coercivity was determined on the basis of the major demagnetization curve *B* vs. *H* as the value equal to the magnetic field strength required for changing the magnetic induction from the remanence to zero.

The maximum energy product was determined on the basis of the major demagnetization curve *B* vs. *H* as the maximum value of the product of the magnetic flux density and magnetic field strength in the second *B*-*H* coordinate quadrant.

3 Results and Discussion

Figure 1 presents an example of the major demagnetization curve for the MS NdFeB-12 from batch No. 001. Analogous curves are constructed for all RMs included in the set. The certified values of the RM characteristics are determined on the basis of the constructed curves. These values were estimated as the arithmetic mean of all measurement results of every value.

The RMs were not studied for heterogeneity, since in establishing the certified values of the RM characteristics, the measuring signal was picked up from the whole RM; thus, the certified value was assigned to the whole RM. Since unit-by-unit RM certification was carried out, it was not necessary to conduct between-unit homogeneity studies.

A stability study was carried out along with a determination of the RM validity period and an estimation of uncertainty due to long-term instability u_{stab} of the

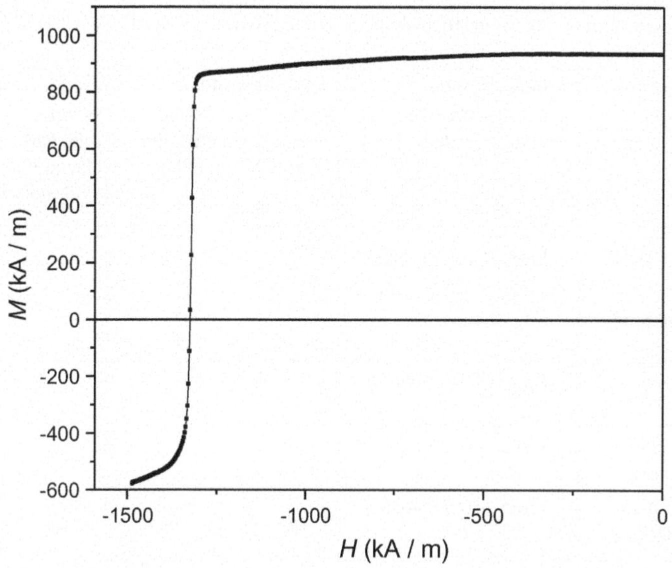

Fig. 1 Major demagnetization curve *M* vs. *H* of RM limit curve (MS NdFeB-12 from batch No. 001)

magnetic properties of NdFeB alloy-based RM material. These studies were carried out using a classical method in accordance with the algorithms given in [2]. Two RM units selected for the studies from the test batch were stored at a temperature of (22 ± 5) °C.

In calculating the expanded uncertainty of the RM certified values, the following is taken into account: uncertainty due to the RM characterization method and uncertainty due to the long-term instability of the RM material. The expanded uncertainty of each certified value was estimated as follows

$$U = 2\sqrt{u_{char}^2 + u_{stab}^2}, \quad (4)$$

where u_{char} is the standard uncertainty due to the RM characterization method; u_{stab} is the standard uncertainty due to RM short-term instability.

The basic requirements for the RM metrological characteristics, established in the RM type description, are given in Table 1.

The sufficiently broad ranges of the permissible values as shown above allow this RM type to be produced with each RM certified characteristic having slightly differing values.

In the process of RM development, it was established that the expanded uncertainty and the bounds for the absolute error of the certified values of the material magnetic properties from any batch do not exceed the values outlined in the RM type description, with due observance of the technical specification requirements for the RM development and the test program when RMs are mass produced.

Table 1 Basic requirements for the metrological characteristics of GSO 11059-2018/GSO 11062-2018

RM certified characteristic, measurement unit	Interval of the permissible RM certified values	The permissible relative error bounds of RM certified value (P = 0.95), ± δ, %	Permissible relative expanded uncertainty of the RM certified value at P = 0.95 and $k = 2$, not exceeding, %
Remanence, B_r,T	From 0.900 to 1.500	2	2
Remanent magnetization, M_r kA/m	From 715.0 to 1200.0	2	2
Normal coercivity, H_{cB}, kA/m	From 680.0 to 990.0	3	3
Intrinsic coercivity, H_{cM}, kA/m	From 800.0 to 3500.0	3	3
Maximum energy product, $(BH)_{max}$, kJ/m^3	From 150.0 to 400.0	4	4

The certified values of remanence, remanent magnetization, normal coercivity, intrinsic coercivity and maximum energy product are traceable:

- to the length unit (m) using a micrometer in the determination of RM metrological characteristics, verified via the uninterrupted verification chain, in accordance with [3];
- to the magnetic induction unit (T) and magnetic flux unit (Wb) by means of direct measurements using the GET 198-2017 State Primary Measurement Standard for the units of magnetic loss power, magnetic flux density of constant magnetic field in a range from 0.1 to 2.5 T and magnetic flux in a range from 1×10^{-5} to 3×10^{-2} Wb.

Documents on the developed RM set are drawn up in accordance with [2, 4].

In order to evaluate the RM applicability, the hysteresis magnetic properties of RMs from batch No. 001 were measured using a Permagraph L (introduced into the State Register under No. 59000-14). The conducted tests showed the fitness of the developed RM set for the metrological support of similar-type magnetic measuring systems.

4 Conclusion

On the basis of the conducted studies, an RM set for the magnetic properties of NdFeB alloy-based hard magnetic material was developed and approved (MS NdFeB set). The developed RMs for the magnetic properties of NdFeB alloy-based hard magnetic material (MS NdFeB set) were introduced into the State RM Register of the approved type of the Russian Federation under the numbers GSO 11059-2018/GSO 11062-2018. The information on the development of GSO 11059-2018/GSO11062-2018 is outlined in the Federal Information Fund on Measurement Uniformity Assurance.

The RMs are intended for calibrating and testing instruments for measuring the magnetic properties of hard magnetic materials, including for the purposes of type approval; certifying measurement procedures and controlling the accuracy of results of measuring the properties of hard magnetic materials; other forms of metrological control provided RM metrological characteristics comply with the metrological control procedural requirements.

The RMs can be used for verifying instruments for measuring the magnetic properties of hard magnetic materials, provided the metrological and technical characteristics of the RMs meet the criteria outlined in the verification procedures of the corresponding measuring instruments.

The NdFeB alloy-based RMs for the magnetic properties of hard magnetic materials are directly traceable to GET 198-2017.

At the moment, work is currently in progress to produce new batches and types of NdFeB alloy-based RMs for the magnetic properties of hard magnetic materials, as well as studying other kinds of hard magnetic materials for the purpose of developing new type approved RMs.

Author Contribution Statement Volegova E. A.: measurement, processing of measurement results, preparation of the initial text version of the paper, text formatting.

Maslova T. I.: scientific management, determination of the concept and methodology of the paper.

Vaskovskiy V. O.: scientific management, critical analysis and text modifications.

Volegov A. S.: measurement, concept and study initiation; critical analysis and text modifications; text formatting; formalized data analysis.

References

1. Volegova EA, Malygin MA, Maslova TI, Volegov AS (2018) GET 198–2017, state primary standard of unit of power of magnetic losses and unit of magnetic induction of constant magnetic field in range from 0.1 to 2.5 T and magnetic flux from $1{\cdot}10^{-5}$ to $3{\cdot}10^{-2}$ Wb. Meas Tech 61(3):199–202. https://doi.org/10.1007/s11018-018-1409-2
2. GOST ISO (2016) Guide 35–2015 reference materials. General and statistical principles for certification. Standartinform, Moscow, p 61 (In Russ.)
3. GOST R 8.763-2011 state system for ensuring the uniformity of measurements. State verification schedule for measuring instruments of length in the range from $1{\cdot}10^{-9}$ to 50 m and wavelengths in the range from 0.2 to 50 μm. Standartinform, Moscow, 2013, 7 p. (In Russ.)
4. GOST R 54500.3–2011 Uncertainty of measurement. Part 3. Guide to the expression of uncertainty in measurement. Standartinform, Moscow, 2012, 107 p. (In Russ.)

Mathematical and Statistical Support of Reference Material Characterisation

Estimation of Consensus Value of Interlaboratory Measurement Results Accompanied by a Minimum Increase in Associated Uncertainty

Peter M. Aronov

Abstract In this article, the author proposed a method for estimating the consensus value of interlaboratory measurement results, which makes it possible to reduce the estimation uncertainty in comparison with the known methods. The method is based on the selection in the inconsistent data of the largest consistent subset of the results, the uncertainties of which do not increase when calculating the consensus value.

Keywords Interlaboratory measurements · Inconsistent data · Consensus value · Uncertainty

The use of interlaboratory experiment in the characterisation of reference materials along with other metrological problems necessitates estimation of the consensus value of the measurand according to measurement results $\{x_i, u_i\}, i = \overline{1, n}$ (measurement result and its standard uncertainty), obtained in n independent laboratories. It is usually assumed that errors follow a normal distribution. In which case, weighted average constitutes the maximum likelihood estimate of the measurand.

$$\bar{x_0} = u_0^2 \sum_{i=1}^{n} u_i^{-2} x_i, \quad (1)$$

where u_0^2 is the square of standard uncertainty (variance) of the estimate (1)

$$u_0^2 = \left(\sum_{i=1}^{n} u_i^{-2} \right)^{-1}, \quad (2)$$

This statistic

$$g = \sum_{i=1}^{n} \frac{(x_i - \bar{x}_0)^2}{u_i^2} \quad (3)$$

P. M. Aronov (✉)
Ural Scientific Research Institute for Metrology (UNIIM), Yekaterinburg, Russia
e-mail: AronovPM@uniim.ru

S. Medvedevskikh et al. (eds.), *Reference Materials in Measurement and Technology*,
https://doi.org/10.1007/978-3-030-32534-3_15

has a chi-square distribution and the following condition of a chi square test must be satisfied

$$g \leq \chi^2(P; n-1), \tag{4}$$

where $\chi^2(P; n-1)$ is the P–quantile for the chi-square distribution with $n-1$ degrees of freedom. In this case, the data $\{x_i, u_i\}, i = \overline{1, n}$ are deemed consistent. If condition (4) is not satisfied, the data are considered to be inconsistent, i.e. measurement results obtained from some laboratories contain systematic errors not included in the stated uncertainties.

Data inconsistency, i.e. the dispersion of measurement results x_i between laboratories is often described by normal distribution centred on the target value of the measurand μ and interlaboratory variance σ^2. In which case, in order to make the data consistent, the standard uncertainties of each laboratory are increased from u_i^2 to $u_i^2 + \sigma^2, i = \overline{1, n}$ followed by calculation of the weighted average $\bar{x}_{\sigma^2}$ as well as its variance $u_{\sigma^2}^2$ according to formulas (1–2) and using new weights.

There are known methods for estimating σ^2, for example, maximum likelihood method, Der Simonian–Laird method or Mandel-Paule algorithm [1]. The latter employs the following equation for calculating σ^2

$$\sum_{i=1}^{n} \frac{(x_i - \bar{x}_{\sigma^2})^2}{u_i^2 + \sigma^2} = n - 1, \tag{5}$$

since the statistic in the left part of (5) with the proper choice of σ^2 value has a chi-square distribution and, therefore, the expected value equal to $n - 1$. Following an increase in the measurement result uncertainties achieved in the above-described way, the data become consistent.

An obvious weakness of this approach lies in a sometimes unjustified increase in the uncertainties of measurement results provided by all laboratories and, as a consequence, an increase in the uncertainty of the target measurand estimate. In this paper, we propose a more economical way of increasing uncertainty to make data consistent. It implies selecting a largest subset of measurement results that are already consistent, therefore it is not necessary to increase their uncertainty.

Let us use the procedure proposed in [2]. If the consistency condition (4) for n laboratories is not satisfied, we arrange the pairs of data $\{x_i, u_i\}, i = \overline{1, n}$ in such a way, that

$$\frac{(x_1 - \bar{x}_0)^2}{u_1^2} < \frac{(x_2 - \bar{x}_0)^2}{u_2^2} < \ldots < \frac{(x_n - \bar{x}_0)^2}{u_n^2}, \tag{6}$$

Then we discard laboratory data for n and recalculate the estimates of (1), (2), as well as chi-square statistic (3) for $n - 1$ laboratories followed by repeating the process until the consistency condition is satisfied for the first time at some k.

$$\sum_{i=1}^{k} \frac{(x_i - \bar{x}_k)^2}{u_i^2} \leq \chi^2(P; k-1), \quad \bar{x}_k = \left(\sum_{i=1}^{k} u_i^{-2}\right)^{-1} \sum_{i=1}^{k} u_i^{-2} x_i, \tag{7}$$

Thus, largest consistent subset k = 2, .., n is selected. In [2] it is proposed to consider the data that are not included in the found subset to be statistical outliers and to discard them, considering $\bar{x}_k$ to be the estimate of consensus value. This is justified if the size of the consistent subset is large and there is a small number of outliers; however, for a general case, this approach seems too wasteful. Instead of discarding the results of laboratories that are not included in the largest consistent subset, we propose to increase their uncertainty to the level sufficient to make them consistent.

Let us consider the parametric family of statistics

$$\bar{x}_\lambda = u_\lambda^2 \left(\sum_{i=1}^{k} u_i^{-2} x_i + \sum_{i=k-1}^{n} \left(u_i^2 + \lambda\right)^{-1} x_i\right),$$

$$u_\lambda^2 = \left(\sum_{i=1}^{k} u_i^{-2} + \sum_{i=k+1}^{n} \left(u_i^2 + \lambda\right)^{-1}\right)^{-1}, \tag{8}$$

$$g(\lambda) = \sum_{i=1}^{k} \frac{(x_i - \bar{x}_\lambda)^2}{u_i^2} + \sum_{i=k+1}^{n} \frac{(x_i - \bar{x}_\lambda)^2}{u_i^2 + \lambda}, \tag{9}$$

where $\lambda \geq 0$ specifies an increase in variances for data that are not included in the largest consistent subset. With a proper choice of λ value, formula (8) provides an estimate of the measurand and its variance, whereas formula (9) produces chi-square statistic, which takes an increase in the variances of the inconsistent data into account. Assuming that the dispersion of measurement results for data not belonging to the largest consistent subset is characterised by a normal distribution with variance σ^2, at $\lambda = \sigma^2$ the statistic (8) has chi-square distribution with $n - 1$ degrees of freedom expected value equal to n − 1. By analogy with the Mandel-Paule algorithm (5), the value of λ can be obtained using the following equation.

$$g(\lambda) = n - 1 \tag{10}$$

However, the difficulty lies in the fact that Eq. (10), unlike Eq. (5), does not always have a solution. Differentiating by λ formula (9), it is easy to see that

$$\frac{\mathrm{d}g(\lambda)}{\mathrm{d}\lambda} = -\sum_{i=k+1}^{n} \frac{(x_i - \bar{x}_\lambda)^2}{\left(u_i^2 + \lambda\right)^2} < 0, \tag{11}$$

Therefore, the function $g(\lambda)$ decreases monotonically with increasing λ

$$\lim_{\lambda\to\infty} g(\lambda) = \sum_{i=1}^{k} \frac{(x_i - \bar{x}_k)^2}{u_i^2}, \tag{12}$$

Hence, taking (7) into account, it follows that Eq. (10) has a solution if and only if

$$\sum_{i=1}^{k} \frac{(x_i - \bar{x}_k)^2}{u_i^2} \leq \chi^2(P; k-1) < n - 1, \tag{13}$$

The fulfilment of condition (13) depends only on the ratio of the number of laboratories n and the size of the largest consistent subset k. At sufficiently small k, the condition is satisfied and the solution of Eq. (10) can easily be found numerically. It is a statistical estimate of interlaboratory variance σ^2. On substituting it into formulas (8), we obtain an analogue of the Mandel-Paule estimate of the consensus value, as well as the square of its standard uncertainty.

If condition (13) is not satisfied and Eq. (10) has no solution, then the required value of λ can be found using the inequality

$$g(\lambda) \leq \chi^2(P; n-1), \tag{14}$$

which in itself is the consistency condition for data from n laboratories. Inequality (14) always has a solution. Let us demonstrate it. From (12) and the consistency condition (7), it follows that for every $\varepsilon > 0$ and sufficiently large λ the following inequality holds

$$g(\lambda) \leq \chi^2(P; k-1) + \varepsilon, \tag{15}$$

From the properties of the chi-square distribution, it follows that $\chi^2(P; k-1) < \chi^2(P; n-1)$ when k < n. Let us choose $\varepsilon = \chi^2(P; n-1) - \chi^2(P; k-1)$ and, by substituting it into (15), we obtain the required inequality (14).

In the case when σ^2 is estimated by solving inequality (14), the Birge factor should be used for the squared uncertainty of the weighted average (8)

$$u_{cor}^2(\bar{x}_{\sigma^2}) = u_{\sigma^2}^2 \frac{g(\sigma^2)}{n-1} \leq u_{\sigma^2}^2 \frac{\chi^2(P; n-1)}{n-1}, \tag{16}$$

Thus, provided the data are consistent (k = n), the proposed algorithm offers a standard solution (1); whereas in case of inconsistent data (1 < k < n), it guarantees smaller uncertainty than the methods recommended in [1] and only in the absence of a consistent set (k $\leq$ 1) is equivalent to them (Mandel-Paule algorithm).

We will illustrate the proposed algorithm by the example of estimating the consensus value in the interlaboratory measurements of higher heating value. The measurements in question were performed by 20 laboratories for a reference material of coal composition. On the graph, the results provided by the laboratories and their expanded uncertainties are arranged in accordance with the algorithm, so that the

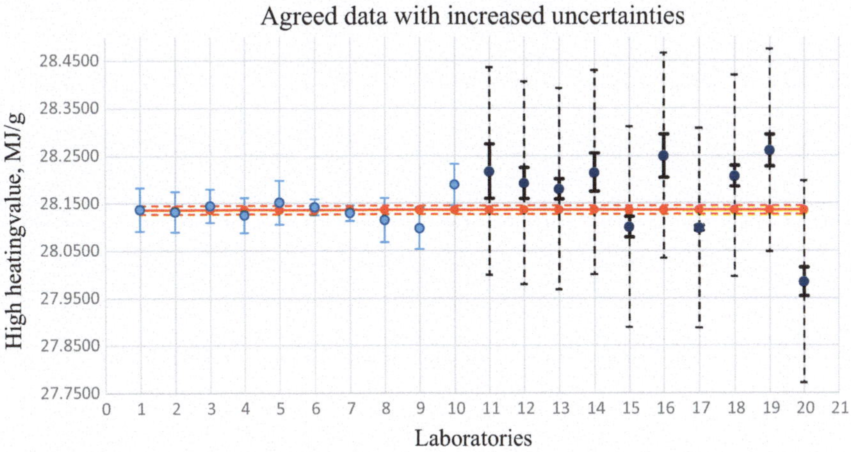

Fig. 1 Reference material of coal composition

left-hand side contains the largest consistent subset. In this case, it consists of 10 laboratories, i.e. half of the experiment participants. In line with the algorithm, the remaining participants are arranged in ascending order of inconsistency of their results. Vertical dashed lines indicate an increase in the measurement result uncertainty expanded for achieving data consistency in accordance with the algorithm. In this case, the estimate of interlaboratory variance σ^2 was found as a solution to Eq. (10), since for $P = 0.95$, $k = 10$, $n = 20$, condition (13) is satisfied

$$\chi^2(0.95; 9) = 16.9 < 19$$

and Eq. (10) has a solution. On the graph, a corresponding consensus value is represented by a horizontal solid line; whereas horizontal dashed lines indicate the bounds of its expanded uncertainty (Fig. 1).

References

1. CCQM Guidance note: Estimation of consensus KCRV and associated Degrees of Equivalence. Version: 10, 12 Apr 2013
2. Cox MG (2007) The evaluation of key comparison data: determining the largest consistent subset. Metrologia 44:187–200

Quality Assessment of the Preparation of Reference Materials of Moist Solid Substances

Sergej V. Medvedevskikh, Maria Yu. Medvedevskikh and Maria P. Krasheninina

Abstract In this article, the models of random parameter and random walk are considered as alternative approaches for estimating the mean standard deviation associated with heterogeneity during both the evaluation of the metrological characteristics of reference materials, as well as the quality control of samples prepared from moist solid substances. For the random parameter and random walk models, the behaviour of the basic statistical criteria is described, including, overall mean, standard deviation, standard deviation of the mean and autocorrelation function, obtained both for real experimental data and theoretical measurement results simulated according to the Monte Carlo method. It is shown that the application of the ANOVA method for evaluating the standard deviation associated with heterogeneity should be additionally validated for the materials, whose basic statistical criteria fall under the random walk model.

Keywords Certified reference material · Processing quality of certified reference material · Homogeneity study of moist solid substances · Monte Carlo method · ANOVA method · GET 173-2017 · Random walk model

1 Introduction

An essential prerequisite for the development of reference materials (RMs) consists in the evaluation of the standard deviation induced by the inhomogeneity of materials used to produce RMs. When investigating RMs in terms of their homogeneity, the ISO Guide 35:2017 "Reference materials—Guidance for characterisation and

Maria Yu. Medvedevskikh—ResearcherID: G-6171-2017
Maria P. Krasheninina—ResearcherID: B-8302-2019

S. V. Medvedevskikh · M. Yu. Medvedevskikh · M. P. Krasheninina (✉)
Ural Scientific Research Institute for Metrology (UNIIM), Ekaterinburg, Russia
e-mail: krasheninina_m@uniim.ru
URL: https://publons.com/researcher/1794033/maria-krasheninina/

M. Yu. Medvedevskikh
URL: https://publons.com/researcher/2119462/mary-medvedevskikh/

S. Medvedevskikh et al. (eds.), *Reference Materials in Measurement and Technology*,
https://doi.org/10.1007/978-3-030-32534-3_16

assessment of homogeneity and stability" [1] recommends the use of the ANOVA method for the analysis of measurement results. This method is capable of processing multiple results obtained during the analysis of the same sample and allows the dispersions associated with heterogeneity and measurement uncertainty to be estimated separately. However, when destructive methods of analysis are used, e.g. in determination of the water mass fraction in solids and solid substances [2, 3], only single (univariate) measurements of the composition characteristics can be obtained, thus requiring additional substantiations for the ANOVA application. Such methods are often used for determining the mass fraction of moisture in solid bulk materials [4–6]. The mass fraction of moisture is known to be one of the most unevenly distributed indicators in terms of volume. Therefore, this indicator has been selected to study the specific features of material processing during the development of RMs of solid substances.

If the measurement results x_i are obtained for RM samples, taken sequentially in time from the total volume of the material selected for the characterisation of an RM, they can be modeled as a random stationary number sequence, for which the expected value, variance and covariance function can be determined. In this case, in order to characterise an RM on the basis of processing the experimental data x_i, where $i = \overline{1, N}$, well-known methods of the random process theory and the methods of consistent statistical analysis can be used.

In the present work, the use of the random parameter (normal distribution with a detectable influencing factor and independent measurement results) and random walk (normal distribution with a detectable influencing factor and dependent measurement results) models is considered for characterisation of RMs of the mass fraction of moisture in solid substances using the thermogravimetric method [7–9]. This approach permits preliminary decisions on the quality of sample preparation and RM homogeneity to be made.

2 Theoretical Part

2.1 *Model of the Moist Dispersed Substance in Case of Sequential Sampling from One Vessel*

Let there be a vessel with the volume V (m^3) and the mass M (kg) containing a moist solid substance. Samples of the substance having the same mass m (kg), the mass fraction of water $x = m_W/m$ (kg/kg), m_W and the mass of water in a sample m_W (kg) are taken from this vessel successively in time. The number of samples is limited, with their total mass being lower than the total mass of the substance in the vessel, M.

The mass fraction of water in each sample is measured by a destructive method with an unknown uncertainty in the general case.

The measurement results of the water mass fraction in the samples can be applied for the following purposes:

1. Assesment of the water mass fraction in a substance with a mass of M.
2. Assesment of the water mass fraction in the substance remaining in the vessel after sampling.
3. Homogeneity study of samples by the mass fraction of water.
4. Quality assessment of the processes of preparation and sampling.
5. Reproducibility assessment of the sampling and preparation processes (i.e. quality of sampling and sample preparation techniques).
6. Evaluation of uncertainties associated with sample inhomogeneity and measurement errors.

In order to formulate a statistical model describing the distribution of water by volume V of the substance and the processes of sampling and sample preparation, we assume that the substance in the vessel can be divided into volumes V_i with a mass $m_i = m$, such that $\sum_{i=1}^{N} m_i = Nm = M$, with the mass fraction of water in the volume V_i equalling $x_i = m_{Wi}/m$, where m_{Wi} is the mass of water in i-volume of the moist substance.

Then the mass fraction of water with the mass of M can be calculated as:

$$\bar{x}_N = \frac{m_W}{M} = \frac{1}{N}\sum_{i=1}^{N} x_i, \tag{1}$$

where $\sum_{i=1}^{N} m_{Wi} = m_W$.

The inhomogeneity in terms of the water mass fraction among the samples with the volumes V_i, where $i = \overline{1, N}$ can be characterised by the dispersion associated with inhomogeneity

$$\sigma_H^2 = \frac{1}{N}\sum_{i=1}^{N} (x_i - \bar{x}_N)^2. \tag{2}$$

According to [10], when taking n-samples, where $n \le N$, the value of $\bar{x}_N$ can be estimated as follows

$$\hat{\bar{x}}_N(n) = \frac{1}{n}\sum_{i=1}^{n} \hat{x}_i, \tag{3}$$

$$D\hat{\bar{x}}_N(n) = \left(1 - \frac{n}{N}\right)\frac{Dx}{n}, \tag{4}$$

where $Dx = 1/(n-1)\sum_{i=1}^{n} (\hat{x}_i - \hat{\bar{x}}_N(n))^2$ and $\hat{x}_i$ are the measurement results of x_i quantities, characterised with unknown uncertainties $u(x_i)$, where $i = \overline{1, n}$, in the general case.

If the x_i values are independent and normally distributed, and multiple independent measurements of the x_i values, where $i = \overline{1, n}$, with the same uncertainty $u(x)$ are available, then the application of the ANOVA method is proposed to estimate the values of σ_H and $u(x)$, according to [10].

The modelled result of measuring the mass fraction of water $\hat{x}_{ij}$ will be as follows

$$\hat{x}_{ij} = \bar{x}_N + \delta_{Hi} + \varepsilon_{ij}, \tag{5}$$

where δ_{Hi} is a random variable characterising a random variation in the values of the mass fraction of water between samples; $\forall i, j$: $\mathrm{E}(\delta_{Hi}) = 0$; $D(\delta_{Hi}) = \sigma_H^2$; $\mathrm{cov}(\delta_{Hi}, \delta_{Hj}) = 0$; $i \neq j$; ε_{ij} is a random variable characterising measurement errors; $\forall i, j$: $\mathrm{E}(\varepsilon_{ij}) = 0$; $D(\varepsilon_i) = \sigma_\varepsilon^2 = u^2(x)$; $\forall i, i', j, j', \mathrm{cov}(\varepsilon_{ij}, \varepsilon_{i'j'}) = 0$, with $i \neq i'$ and $j \neq j'$; $\mathrm{cov}(\delta_i, \varepsilon_{i'j}) = 0$, $\mathrm{E}(\delta_{Hi}) = 0$; E(.), D(.) and cov(., .), denote the expected value, variance, and correlation of random variables, respectively; $i, i' = \overline{1, n}$, $j, j' = \overline{1, J}$, where J is the number of measurements of the x_i values.

In present work, we consider the situation, when the number of analysed samples $n \ll N$, with each sample being analysed once. The value of N is unknown. In this case, within the framework of the approach [1], σ_H and $u(x)$ values are impossible to be determined separately.

However, if the values x_i and $\hat{x}_i$ are represented as follows

$$x_i = \bar{x}_N + \delta_{Hi}, \tag{6}$$

$$\hat{x}_i = x_i + \varepsilon_i, \tag{7}$$

then, when the conditions of applicability of ANOVA for $J = 1$ specified in the present work for model (5) are fulfilled for model (6)–(7), the estimate of $\bar{x}_N$ will be presented by the estimate obtained by formula (3) with the following variance

$$D\hat{\bar{x}}_N(n) = \sigma_H^2 + \frac{1}{n}\sigma_\varepsilon^2. \tag{8}$$

Sometimes [2], model (6)–(7) is called the random parameter model [11].

If N is unknown, then the variance of the value $\hat{\bar{x}}_N(n)$ can be estimated from the $\hat{x}_i$ measurement results in the form of $D\hat{\bar{x}}_N(n) = \frac{1}{n(n-1)} \sum_{i=1}^{n} (\hat{\bar{x}}_N(n) - \hat{x}_i)^2$. Equation (8) can be considered as a linear regression equation for the unknown coefficients of σ_H^2 and σ_ε^2 represented by following formula

$$Dx(j) = \sigma_\varepsilon^2 + j \cdot \sigma_H^2, \tag{9}$$

where $Dx(j) = \frac{1}{j-1} \sum_{k=1}^{j} (\bar{x}_j - x_k)^2$ is the current variance, $\bar{x}_j = \frac{1}{j} \sum_{k=1}^{j} x_k$ is the current mean with $j = \overline{2, n}$. It may turn out that the value of σ_H^2 obtained from (9) is large and does not suit the experimenter.

Then, before sampling, the substance in the vessel is mechanically averaged in terms of the mass fraction of water by mixing.

Subsequently, in order to build a statistical model describing changes in the x_i quantities after mixing the substance, the following assumptions should be fulfilled.

1. The substance is a macroscopic system consisting of the following components: solid particles forming a porous or dispersed solid base; water molecules both inside and on the surface of the solid base in the liquid, ice or vapour form, or as part of the solid base in a structurally bound (hydrated) form in the case when the substance of the solid base is a crystalline hydrate; other non-water compositional components distributed both inside and on the surface of the solid base as liquid, gas or the solid base dissolved in water and present inside or on the surface of the solid base. This means that moist substances in the form of gases, organic liquids and suspensions are excluded from consideration by this model, and the constancy of the chemical composition of the moist solid substance and thermodynamic limiting period are assumed. In other words, the system is macroscopic and can be characterised by the mean values of the microscopic densities of the system particles.
2. The measurement object is a moist solid substance with the volume $V(t)$ and the mass of $m(t)$ at the time of measurement $t \in [t_0, T]$ and, during T measurement time period, is characterised by an equilibrium state resulting from irreversible processes occurring during the formation or production of a solid substance. This implies the absence of chemical reactions and gradients of mean densities (concentrations), partial pressures of liquids and gases in the composition of the moist solid substance and the temperature at the time point t. Thus, we can assume that $t \gg t_p$, where t_p is the relaxation time of the measurement object up to the equilibrium state.
3. The measurement object is a quasi-closed system that remains in thermal and material contact with the ambient air environment, with the gradients of mean parameters (temperature, pressure and humidity) being absent during the measurement time period T and the preparation period before the measurements τ. Mass exchange is carried out only through the exchange of water molecules in the form of vapour. This implies that, in periods of time T and τ, the mass of the dry basis of a moist solid is considered to be a constant non-random value, while the water mass may be a random value as a result of the presence of thermodynamic fluctuations due to mass exchange with the environment. This, in turn, assumes small, but comparable in magnitude with measurement errors, fluctuations of the values of environmental parameters and the mass of water in a solid substance compared with its mean values for periods of time T and τ. At the same time, T and τ should be much longer than the correlation time presented by rapid changes in the values that bring the system to a quasi-equilibrium state.
4. The object of measurement is a moist solid substance, a spatially quasi-homogeneous system with a characteristic macroscopic size of $L \gg r_c$, where r_c is the radius of correlations of the microscopic densities of the particle number in the system. This implies that the macroscopic system of particles that make

up the measurement object can be divided into quasi-closed and, in the general case, internally-spatially inhomogeneous macroscopic subsystems of volume V_y with a characteristic size $L_y : r_c \ll L_y < L$, in which the mean density values of the number of particles of a substance will differ insignificantly from the mean densities of particle number of a substance of volume V over a period of time T_y. This means that a limited number of measurements of the moisture mass fraction in each of these subsystems $x(V_y, T_y)$ can be used to assess the value of the moisture mass fraction in the system as a whole. In addition, the procedure for selecting such subsystems from the system of volume V is possible, technically feasible and controllable with small relative deviations of the sampling parameters leading to small deviations of $x(V_y, T_y)$ values.

5. An assumption is made that, under the influence of external factors, each subsystem of substance particles with the volume V_y can be transformed into a spatially homogeneous system with masses $m_i = m$ and volumes $V_i = V_\beta \leq V_\gamma$ with a characteristic size $L_\beta : r_c \ll L_\beta < L_y < L$, the density value in which will differ insignificantly from the values in volume V_y.

This implies that, for a period of time T_β, the technical feasibility and controllability of the procedures for preparing a homogeneous system of particles of a volume V_β from the volume V_y is possible under the condition of small relative deviations of the sample preparation parameters leading to small deviations of the mass fraction of moisture in the substance $x(V_\beta, T_\beta)$, with the sample preparation itself ensuring the insignificance of deviations between $x(V_\beta, T_\beta)$ and $x(V_y, T_y)$ values.

Further, it becomes possible to obtain the results of measuring the mass fraction of water $\hat{x}_i$, when $i = \overline{1, n} < N$, for each of the selected samples that underwent conversion during sampling and sample preparation and, in the general case, do not completely cover the specified volume of substance $V(t)$.

It follows from assumptions 4–5 that the behaviour of the mass fraction of water in different substance volumes selected and prepared to estimate the value of the water mass fraction of a given volume and mass, including different time periods, is determined by the similar behaviour of macroscopic values of dry basis density, porosity and the degree of filling of the pores of the moist dry matter.

Then the mass fraction of water in the samples after a single mixing x_{mi}, where $i = \overline{1, N}$, can be presented in a linear form

$$x_{mi} = \sum_{j=1}^{N} \beta_{ij} x_j, \tag{10}$$

where β_{ij} are the coefficients characterising the fraction of the substance from the i-sample transferred after mixing to j-volume of the substance in the case of $i \neq j$; β_{ii} is the fraction of the substance remaining in i-volume of the substance in the vessel after mixing, where $i, j = \overline{1, N}$. Values x_{mi}, where $i = \overline{1, N}$, form a finite set of mass fraction values of water in the vessel after mixing in V_i volumes.

The coefficients β_{ij} are generally unknown; however, if mixing is performed without loss of water in the substance, then $\bar{x}_{mN} = 1/N \sum_{i=1}^{N} x_{mi} = \bar{x}_N$ is fulfilled, and it can be shown that $\sum_{i=1}^{N} \beta_{ij} = \sum_{j=1}^{N} \beta_{ij} = 1$ and $1/N \sum_{i,j=1}^{N} \beta_{ij} = 1$. Note that from (10) and taking into account (6) and the equation $(1 - \beta_{ii}) = \sum_{j=1, i \neq j}^{N} \beta_{ij}$, it can be obtained that $\forall i = \overline{1, N}$

$$x_{mi} = \beta_{ii} x_i + (1 - \beta_{ii}) \bar{x}_N + \gamma_{mi}(\delta_H), \tag{11}$$

where $\gamma_{mi}(\delta_H) \equiv \sum_{j=1, j \neq i}^{N} \delta_{Hj} \beta_{ij}$ are random variables with the expected value $E(\gamma_{mi}(\delta_H)) = 0$.

Let us consider the behaviour of Eq. (11) in two extreme cases. When mixing is perfect $\beta_{ii} = \beta_{ij} = \frac{1}{N}$ and $\forall i = \overline{1, N}$, $x_{mi} = \bar{x}_N$ takes place. From Eq. (11) with the consideration of (6), when $E(\delta(\beta_{ij})) = 0$, for small deviations $\delta(\beta_{ij})$ between β_{ij} and the value $\beta_0 = 1/N$ we obtain

$$x_{mi} = \beta_0 x_i + (1 - \beta_0) \bar{x}_N + \gamma_{mi}(\delta_H, \beta_0), \tag{12}$$

where $\gamma_{mi}(\delta_H, \beta_0) \equiv \bar{x}_N \delta(\beta_{ii}) + \sum_{j=1, j \neq i}^{N} \delta_{Hj} \delta(\beta_{ij})$ are random variables with the expectation $E(\gamma_{mi}(\delta_H, \beta_0)) = 0$ with $E(x_{mi}) = \bar{x}_N$. In other words, $\bar{x}_{mn} = \frac{1}{n} \sum_{j=1}^{n} x_{mj}$ can be considered as an estimate of $\bar{x}_N$, where the number of samples taken after a single mixing is $n \ll N$.

Another extreme case is when the mixing effect is absent. Then $\beta_{ii} = \beta = 1$, $\beta_{ij} = 0$, with $i \neq j$ and $x_{mi} = x_i$. If the effects of mixing are small, then $\beta_{ii} = 1 + \delta(\beta_{ii})$ and $\beta_{ij} = \delta(\beta_{ij})$ with $i \neq j$ and $E(\delta(\beta_{ij})) = 0$, and (11) can be represented as

$$x_{mi} = x_i + \gamma_{mi}(\delta_H, \beta), \tag{13}$$

where $\gamma_{mi}(\delta_H, \beta) \equiv \bar{x}_N \delta(\beta_{ii}) + \sum_{j=1, j \neq i}^{N} \delta_{Hj} \delta(\beta_{ij})$ are random variables with the expected value $E(\gamma_{mi}(\delta_H, \beta)) = 0$.

In this case, discrepancies in the mass fraction of water in the samples may increase. This is the case when mixing may increase the variance due to the heterogeneity of the samples σ_H^2.

In the case considered in this work, samples are taken sequentially, with a single mixing preceeding each sampling. Only one sample is taken from the same place of the vessel with a mass of $m_k = m$, where the k-number indicates the location of the sample of volume V_k in the vessel. A total of $L \ll N$ samples are taken. Thus, the sequence $\{x(t_l)\}_{\overline{l=1,L}}$ is obtained, representing a time series, where $x(t_l) \equiv x_k(t_l)$, $t_0 < t_1 < \cdots < t_L$. After measuring the mass fraction of water in each sample, a sequence of measurement results is obtained $\{\hat{x}(t_l)\}_{\overline{l=1,L}}$. Equation (10) in this case can be rewritten as

$$x(t_l) = \sum_{j=1}^{N-l+1} \beta_{kj}(t_l)x_j(t_{l-1}), \tag{14}$$

where $x_j(t_{l-1})$ is a finite set of values of the mass fraction of water in the vessel after $(l-1)$-th mixing in volumes V_j. For the k-number, the condition $1 \le k \le N-l+1$ is always fulfilled. The normalisation conditions in (14) are $\sum_{i=1}^{N-l+1} \beta_{ij} = \sum_{j=1}^{N-l+1} \beta_{ij} = 1$ and $1/(N-l+1)\sum_{i,j=1}^{N-l+1} \beta_{ij}(t_l) = 1$.

Hereafter, let us assume that the process of mixing and sampling of a substance from the vessel is reproduced. Then $\forall l = \overline{1, L}$, $\beta_{ij}(t_l) = \beta_{ij} + \delta(\beta_{ij}(t_l))$ and $\forall k$, $1 \le k \le N-l+1$, $\beta_{kk}^{l-1} = \beta + \delta(\beta_{kk}(t_l))$ and $E(\delta(\beta_{ij}(t_l))) = 0$ is fulfilled. Given that the procedure of mixing the substance in the vessel and sampling is aimed at obtaining homogeneous samples, the processing of the results of measurements of the water mass fraction will ensure the achievement of the objectives set at the beginning of this work. Let us assume that model (6)–(7) in this case $\forall k$ can be represented as

$$x(t_l) = \bar{x}_N + \delta_H(t_l), \tag{15}$$

$$\hat{x}(t_l) = x(t_l) + \varepsilon(t_l). \tag{16}$$

The expected values of the random variables $\delta_H(t_l)$ and $\varepsilon(t_l)$ are considered to be equal to zero.

Then from the Eq. (14), taking into account (15) and (16), under the assumption of the reproducibility of the mixing process and sampling, and following the reasoning when obtaining formula (11), we can obtain

$$\hat{x}(t_l) = \beta x(t_{l-1}) + (1-\beta)\bar{x}_N + \gamma(t_l, \delta_H, \delta_\beta, \varepsilon), \tag{17}$$

where $\gamma(t_l, \delta_H, \delta_\beta, \varepsilon)$ is a random variable due to the heterogeneity of the substance in terms of the mass fraction of water and due to random deviations of the processes of mixing and sampling of the substance from the vessel from the conditions established by the experimenter and measurement errors with expected value $E(\gamma(t_l, \delta_H, \delta_\beta, \varepsilon)) = 0$.

In the general case, expression (17) is a 1st order biased autoregression model. Therefore, to test the statistical hypothesis of the stationarity of the sequence $\{\hat{x}(t_l)\}_{l=\overline{1,L}}$, the Dickey-Fuller test [12] in the general version under the assumption that $\forall l = \overline{1, L}$, $\Delta t_l = t_l - t_{l-1} = \Delta t$ can be used.

If the hypothesis of stationarity of the sequence $\{\hat{x}(l)\}_{l=\overline{1,L}}$ is confirmed, and the trend is absent, then, having determined the length of the correlations between $\hat{x}(t_1)$, an expression (9) can be applied to estimate σ_H and σ_ε. In this case, $\bar{x}_L = \frac{1}{L}\sum_{j=1}^{L} \hat{x}(t_j)$ presents an estimate of $\bar{x}_N$.

If the hypothesis of stationarity of the sequence $\{\hat{x}(l)\}_{l=\overline{1,L}}$ is not confirmed, then the quality of sample preparation (mixing) and sampling is considered unsatisfactory.

However, even in this case, if the sequence $\{\hat{x}_i(l)\}_{l=\overline{1,L}}$ can be described **by the model of random walk** [11], when $\beta = 1$ in formula (17), then, using sampled values of the variogram for the time series $\{\hat{x}(t_l)\}_{l=\overline{1,L}}$, σ_H and σ_ε can be estimated separately. This corresponds to the case considered earlier, where the effect of decreasing the σ_ε value after mixing may be absent. On the other hand, depending on the experimenter's objectives, the estimates of σ_H and σ_ε, obtained in terms of the model (8) and (9), may be useful. In addition, the data obtained can be used to establish the optimal number of parallel measurements.

The main difference between the models of random parameter and random walk consist in fact that the measured quantity (x_i) depends on the previous result, and the variance from heterogeneity (D_h) is calculated using the Kolmogorov structure function.

The study of the behaviour of the main statistical criteria was carried out using modelling of the measurement process by the Monte Carlo method [13]: possible measurement processes (a), view of one of the possible processes and its overall mean (b), standard deviation and standard deviation of the mean (c), autocorrelation functions (d) for the normal distribution, model of random parameter and random walk model—and presented in Fig. 1. Figure 1 shows that, for the normal distribution, the measurement results are random, the standard deviation induced by inhomogeneity is absent, the standard deviation of a single measurement result is fairly stable, the standard deviation of the mean decreases with increasing number of measurement results and the autocorrelation function shows no trend. The model of random parameter is characterised by similar behaviour of the main statistical criteria except for the presence of an influencing factor in the form of variance from homogeneity study. However, for the random walk model, the following changes are observed: an increase in the standard deviation of a single measurement and autocorrelation function trend, i.e. the dependence of measurement results.

Upon completion of the theoretical and model studies, the behaviour of the considered models was considered using the results of real experimental data of measuring the water mass fraction.

2.2 Equipment

In present work, in order to empirically verify the considered models for RMs of mass fraction of moisture in solids, the GET 173-2017 State Primary Measurement Standard of units of mass fraction, mass (molar) water concentration in solid and liquid substances and materials [14] (hereinafter referred to as GET 173) was used.

The GET 173 consists of two measurement blocks.

The certified complex implementing the thermogravimetric method for determining the mass fraction of water comprises:

- *Certified instrument of air-heat drying to reproduce a unit of mass fraction of water*;

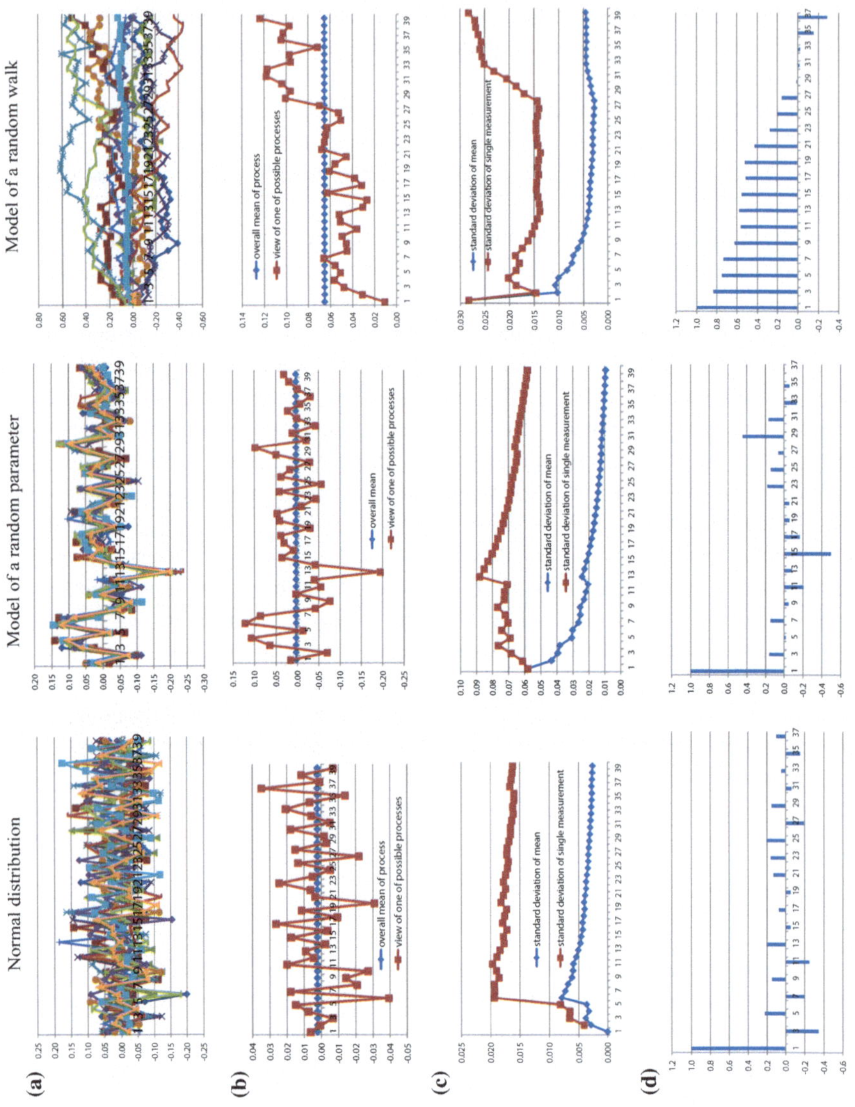

Fig. 1 Illustration of processing and interpretation of measurement results of the water mass fraction, described by the proposed models

- *Certified instrument of vacuum-heat drying to reproduce a unit of mass fraction of water;*
- *Certified instrument of inert gas drying to reproduce a unit of mass fraction of water.*

The certified complex, which implements the physical-chemical methods for determining water, comprises

- *Certified Karl Fischer coulometric titration instrument with drying oven for reproducing units of mass fraction and mass (molar) concentration of water.*
- *Certified Karl Fischer volumetric titration instrument for reproducing units of mass fraction and mass (molar) concentration of water.*
- *Certified instrument of high-temperature coulometry in a thin film of phosphorus pentoxide for reproducing units of water mass fraction.*
- *Certified instrument of thermogravimetric analysis with mass-spectrometric detection, implementing TGA/DSC MS methods.*

A general view of the described certified instruments included in the GET 173 is presented in Fig. 2.

Metrological characteristics of the GET 173-2017 are presented in Table 1.

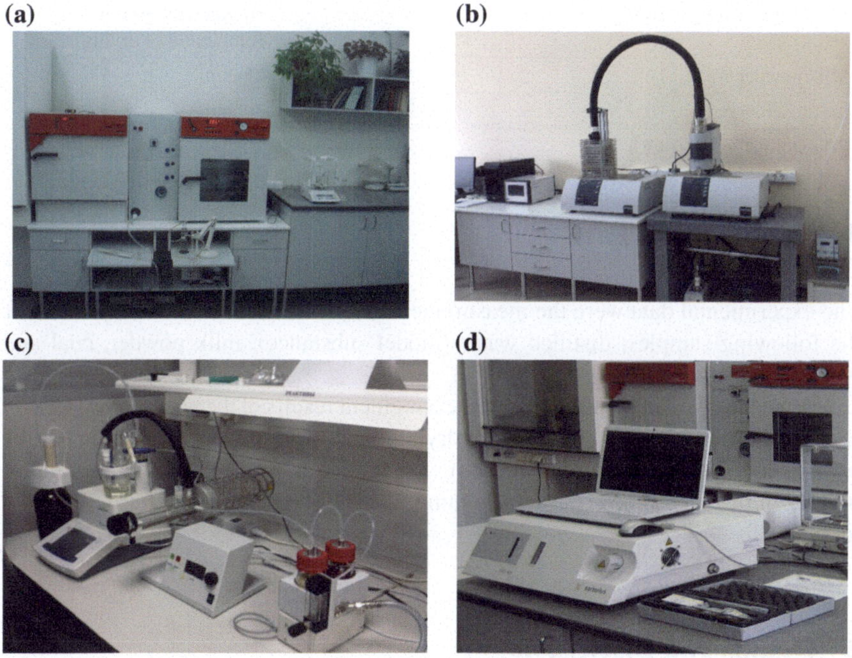

Fig. 2 General view of the GET 173-2017. **a** The certified complex of the thermogravimetric method; **b** certified instrument for thermogravimetric analysis with mass-spectrometric detection; **c** Karl Fischer coulometric titration instrument equipped with a drying oven; **d** certified instrument of high-temperature coulometry in a thin film of phosphorus pentoxide

Table 1 Metrological characteristics of the GET 173-2017 State Primary Measurement Standard of units of mass fraction, mass (molar) concentration of water in solid and liquid substances and materials

Characteristic	Range	U_{Ao}	U_{Bo}	S_0	θ_0
Mass fraction of water (%)	0.001–0.1	3.0–0.6	2.0–0.6	3.0–0.6	4.0–1.2
	0.1–100.0	0.6–0.01	0.6–0.015	0.6–0.01	1.5–0.04
Mass concentration of water (kg/m^3)	0.05–1.0	3.0–0.6	2.0–0.6	3.0–0.6	3.0–1.2
	1.0–900	0.6–0.01	0.6–0.2	0.6–0.01	1.5–0.5
Molar concentration of water (mol/dm^3)	2.0×10^{-3}–0.05	3.0–0.6	2.0–0.6	3.0–0.6	3.2–1.2
	0.05–55.5	0.6–0.2	0.6–0.2	0.6–0.2	1.2–0.4

Designations in the table
S_0—relative standard deviation of the measurement result
θ_0—relative residual systematic error
U_{Ao}—relative standard uncertainty of A type
U_{Bo}—relative standard uncertainty of B type

Table 2 Sample drying modes of the certified air-heat drying instrument included in the GET 173-2017

Sample	Drying temperature (°C)	Weighing (g)	Drying time
Model substance	105	5	30 min with 30 min completion of drying
Milk powder	125	5	25 min
Lignite charcoal	105	2	30 min
Wheat flour	130	5	40 min

3 Experimental Part

The experimental data were the measurement results of the water mass fraction for the following samples: distilled water (model substance), milk powder, coal and wheat flour.

For each of the samples, a series of measurement results of the water mass fraction was obtained using a certified air-heat drying instrument. The measurement results were obtained sequentially in a uniform way. A sample was mixed, and a weighing was taken from the central part of the sample material. Drying modes are presented in Table 2.

4 Results and Discussion

Figures 3 and 4 present the results of processing the experimental data when determining the mass fraction of water in the samples of model substance, milk powder, coal and wheat flour (a—the measurement results and their means; b—the behaviour

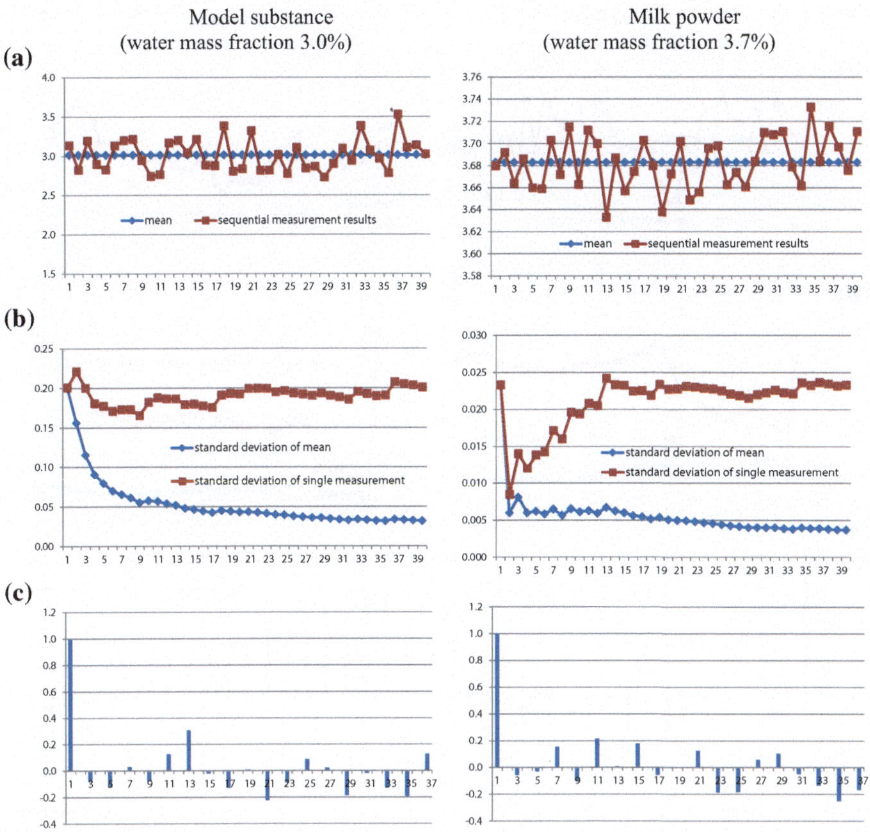

Fig. 3 Processing and interpretation of the measurement data on the water mass fraction based on experimental data for model substance and milk powder

of standard deviation, standard deviation of the mean; c—form of autocorrelation function).

A preliminary analysis of the dependences shown in Figs. 3 and 4 indicates that the sequence of the results of the analysed model substance and milk powder can be described by the random parameter model, while the sequence of the results for wheat flour and lignite charcoal fits the random walk model. Confirmation of the hypothesis about the stationarity of the series of experimental data for a model substance, milk powder and zinc according to the Dickey-Fuller test indicates a satisfactory quality of sample preparation of these materials. At the same time, the experimental data obtained for coal, which are not subject to the conditions of the stationary sequence, indicate the need to refine the conditions for the sample preparation.

Additionally, in order to verify the adequacy of the standard deviation evaluation associated with heterogeneity and the standard deviation from the applied measurement method for the measurement results, which can be described by the

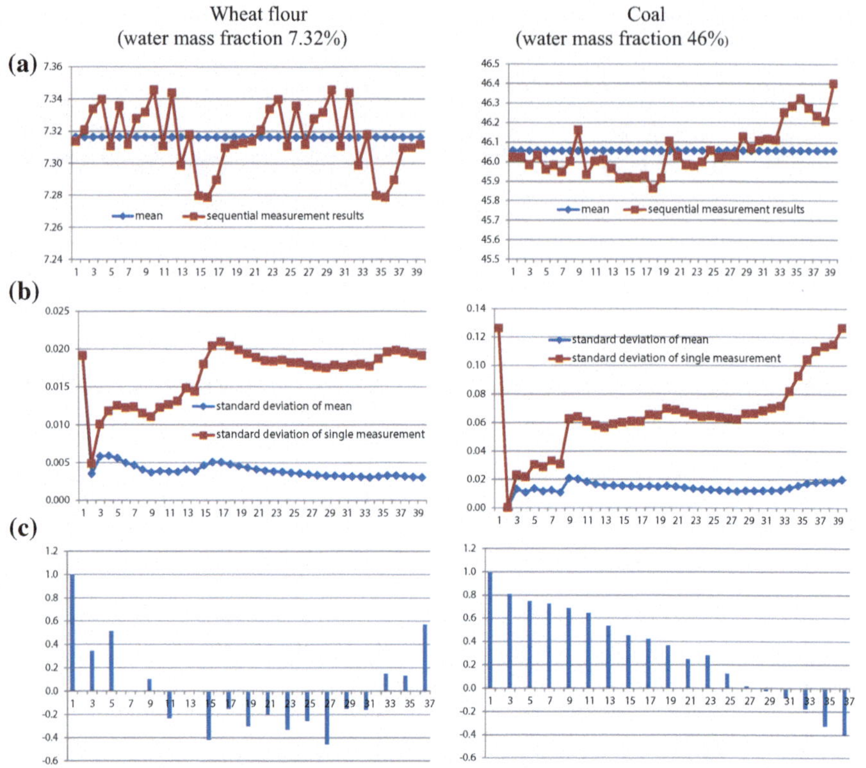

Fig. 4 Processing and interpretation of the measurement data on the water mass fraction based on experimental data for wheat flour and coal

models under consideration, the modelling of the measurement results of the water mass fraction by the Monte Carlo method was carried out [13]. The parameters set in the modelling, as well as the obtained estimates of the standard deviation from homogeneity study and the standard deviation of the method are presented in Table 3.

Table 3 Summary data on the determination of the precision characteristics of modelled measurement results by the proposed methods and by the ANOVA method

Data sets	ANOVA calculated results		Results of calculation using the Kolmogorov function		Parameters specified for Monte Carlo modelling	
	SD_{hom}	SD_{method}	SD_{hom}	SD_{method}	SD_{hom}	SD_{method}
Data set corresponding to a random parameter model	0.045	0.026	0.031	0.020	0.050	0.025
Data set corresponding to a random walk model	0.090	0.057	0.050	0.021	0.050	0.025

Table 4 Summary data on the determination of the precision characteristics of the experimental data by the proposed method and the ANOVA method

Material (substance)—candidate for CRM	ANOVA		Model of random walk	
	SD_{hom}	SD_{method}	SD_{hom}	SD_{method}
Wheat flour	0.012	0.015	0.011	0.013
Coal	0.110	0.060	0.050	0.050

The data obtained show (Figs. 1 and 4) that, in the case of the presence of effects described by the random walk model, the values of the current standard deviation of measurement results increase under an increase in the number of experiments. Using the ANOVA method for such a set of measurement results, conservative values of the SD characteristics (Table 3) were obtained, which may lead to characterisation errors (i.e. incorrect evaluation of the overall mean and material homogeneity).

Tables 3 and 4 present the SD results characterising the homogeneity (SD_{hom}) and those due to the applied measurement method (SD_{method}).

After that, the possibility of applying the proposed approach for processing and interpreting the experimental data was confirmed. Data on the measurements of the water mass fraction in the materials for the development of RMs, including wheat flour and coal, were used. At the stage of assessing the homogeneity of the material and establishing the certified value, five series of eight sequential measurements were performed using the thermogravimetric method. SD results from the homogeneity study and due to the applied measurement method are presented in Table 4.

The data obtained (Table 4) show that the measurement results of the water mass fraction in coal are unsuitable for characterisation, since the effects of correlation of single results were revealed. These effects are likely to results from a deficiency in the analysed sampling weight taken from the prepared total mass, which in turn led to the accumulation of errors and the growth of standard deviation from heterogeneity.

The results obtained on wheat flour can be processed using both the random walk method and the ANOVA single-factor analysis, which is an evidence of the correctness of sample preparation and sample selection procedures, as well as of the correct evaluation of the standard deviation induced by the sample inhomogeneity.

5 Conclusion

1. In this article, the models of random parameter and random walk are considered as alternative approaches for estimating the mean standard deviation associated with heterogeneity when evaluating the metrological characteristics of certified reference materials, as well as for controlling the quality of sample preparation.
2. The behaviour of the total mean, standard deviation, standard deviation of the mean and autocorrelation functions for the random parameter and random walk

models is studied using theoretical (modelled) data. The obtained dependences are confirmed by the results of experimental data during determination of the water mass fraction in solid moist substances. It is shown that an additional substantiation of the ANOVA method is required for evaluation of the standard deviation associated with heterogeneity in cases, where the effects described by the random walk model, are present. Such effects often manifest themselves when measurements are performed by destructive methods of analysis or when only single (univariate) measurement of the characteristics of the substance composition can be obtained.

3. The proposed approach can be applied in the interpretation of the measurement results by initially assessing the behavior of the current dispersion and the autocorrelation function of a series of the measurement results. If you suspect that the data of the random parameter model do not match, it is necessary to test the hypothesis of stationarity of a series of the measurement results based on the application of the Dickey-Fuller test. The proposed approach can be applied in the development of statistical criteria for verifying the quality of RM sample preparation and algorithms for processing measurement information, as well as in the process of characterisation of RMs of heterogeneous materials.

Author Contribution Statement Medvedevskikh S. V.: Concept and methodology of the research, data analysis. Literature review, data curation. Literature review of sources published in foreign languages.

Medvedevskikh M. Yu.: Concept and initiation of the research. Critical analysis and revision of the text. Literature review of sources published in foreign languages.

Krasheninina M. P.: Preparation of the manuscript draft, computer work with the text, preparation of the final text. Collection of experimental data.

Conflict of Interest One of the authors is a member of the Editorial Board of these proceedings.

References

1. ISO Guide 35-2017 Reference materials—guidance for characterization and assessment of homogeneity and stability
2. Medvedevskikh SV, Medvedevskikh MYu, Krasheninina MP (2018) Features of processing the results of single measurements when characterizing standard samples. In: The collection of reports of the international scientific and technical conference. International seminar mathematics, statistics and computation to support measurement quality, 29–31 May 2018. D.I. Mendeleyev Institute for Metrology, St. Petersburg (in Russ.)
3. Medvedevskikh SV, Krasheninina MP, Medvedevskikh MYu (2018) Estimation of material preparation quality of certified reference materials of solid wet substances. In: Conference abstracts III international scientific conference "reference materials in measurement and technology". 11–14 Sept 2018. UNIIM, Yekaterinburg (in Russ.)
4. Medvedevskikh MYu, Sergeeva AS, Krasheninina MP (2015) Certified reference materials in moisture measurements (Review). Ind Lab 81(6):66–71 (in Russ.)

5. Medvedevskikh MYu, Kazennova NK, Krasheninina MP (2013) Certification of measurement techniques and assessment of the applicability of standard samples of the mass fraction of moisture for the control of the quality of food products during their production, processing and storage. Ref Mater (4):40–45 (in Russ.)
6. Krasheninina MP, Medvedevskikh MYu, Medvedevskikh SV, Neudachina LK, Sobina EP (2013) Evaluation of the metrological characteristics of the certified reference material of the dry milk composition using the primary and secondary state standards. Meas Tech 9:67–71 (in Russ.)
7. Sirenek VA (2014) Link relaxation equations mass transfer models "Random walks" and ways of solving diffusion of tasks on their basis. Bull St PbSIT (TU) 49(23):93–96 (in Russ.)
8. Tiannikova ND, Timonin VI (2014) The method of calculating the exact distributions of the Kolmogorov-Smirnov statistics in case of violation of homogeneity and independence of the analyzed samples. Sci Educ Bauman MSTU 11:227–237 (in Russ.)
9. van der Veen AMH, Thomas Linsinger JP (2001) Uncertainty calculations in the certification of reference materials. Homogeneity study. Accred Qual Assur 26–30
10. ISO 11648-1:2003(E) Statistical aspects of sampling from bulk materials—part 1: general principles
11. Mille BM, Pankov AR (2002) The theory of random processes. FIZMATLIT, Moscow (in Russ.)
12. Dickey DA, Fuller WA (1979) Distribution of the estimators for autoregressive time series with a unit root. J Amer Stat 74:427–431
13. ISO/IEC Guide 98-3:2008/Suppl 1:2008. Uncertainty of measurement—Part 3: guide to the expression of uncertainty in measurement (GUM:1995)—Supplement 1: propagation of distributions using a Monte Carlo method
14. Medvedevskikh MYu, Krasheninina MP, Sergeeva AS, Shokhina OS (2018) A reference installation based on thermo-gravimetric analysis with mass-spectrometric detection as a part of the state primary standard GET. Ind Lab 84(6):63–69. https://doi.org/10.26896/1028-6861-2018-84-6-63-68 (in Russ.)

A Method for Estimating the Certified Value Uncertainty of a Multicomponent Reference Material

Pavel V. Migal, Sergei V. Medvedevskikh and Valery A. Firsanov

Abstract This paper presents the results of a series of studies aimed at investigating the reference material (RM) of a multicomponent solution. An algorithm for estimating the uncertainty induced by the chemical element content in the RM is proposed taking into account the mass of the measurand in each component of the mixture. The results calculated according to the preparation procedure using the proposed algorithm were confirmed by the Monte Carlo method, while the elemental content in the mixture was confirmed via inductively coupled plasma optical emission spectroscopy. The applicability of weighted mean estimates for characterizing an RM of a multicomponent solution is shown.

Keywords Multicomponent reference material · Uncertainty · Pure substances · Traceability

Abbreviations

RM	Reference material
GUM	Guide to the expression of uncertainty in measurement
ICP-MS	Inductively coupled plasma mass spectrometry
CGHE	Carrier gas hot extraction
MMC	Monte Carlo method
ICP OES	Inductively coupled plasma optical emission spectrometry

P. V. Migal (✉) · S. V. Medvedevskikh · V. A. Firsanov
Ural Scientific Research Institute for Metrology (FGUP UNIIM), Ekaterinburg, Russia
e-mail: mig@uniim.ru

S. Medvedevskikh et al. (eds.), *Reference Materials in Measurement and Technology*,
https://doi.org/10.1007/978-3-030-32534-3_17

1 Introduction

It is well-known that reference materials (RMs) perform the role of measurement standards in chemistry. RMs of chemical element solutions have been widely used in modern quantitative chemical analysis in constructing calibration dependences of the output signals of, e.g., spectral measurement instruments on the amount of substance. The ease-of-use of RM multicomponent solutions is characterized by the versatility of modern instrumental methods of analysis, which allows calibration of measurement instruments for simultaneous determination of several elements by one or several signals.

2 Literature Review

Over the past eight years, the Ural Scientific Research Institute for Metrology (UNIIM) has accumulated a repository of pure substances, comprising over 70 compounds of 55 individual elements. Having undergone a full purity assessment, these substances are candidates for comparison measurement standards [1]. Such comparison measurement standards in international practice are called *primary reference materials* [2], *primary calibration substances* (*materials*) [3], *primary standards for elemental determination* [4], etc. The work [5] shows that high-purity materials may act as the material realization of a mole unit in the International System of Units (SI); their gravimetrically prepared solutions are primary calibration solutions and can be used in a variety of methods for quantitative chemical analysis. Having a sufficiently large collection of pure substances, it is reasonable to use their solutions (*primary calibration solution* [5]) to characterize the RMs of element solutions (*secondary calibration solution* [5]) using high-performance inductively-coupled plasma spectrometry with high precision [6]. In this case, considering that the relative expanded uncertainty of a mass fraction of the basic component in a solution of comparison measurement standard will be in the order of 0.05%, the relative expanded uncertainty of this component mass fraction in the RM will be in the order of (0.1–0.5)% depending on its stability or the assigned shelf life. Thus, a scheme for ensuring the traceability of RMs of mono-element solutions can be implemented.

As for the production of an RM of a multicomponent solution, it is apparent that the simplest way is to mix mono-element solutions with a well-known analyte content. For the estimation of the uncertainty induced by the preparation method, a method set out in ISO Guide 35: 2017 and based on the use of weighted-mean estimates [7, 9.3.4] is proposed, where the content of a measured component in a mixture is determined using the formulas:

$$\bar{x}_i = \sum_{j=1}^{N} w_j \cdot x_{ij}, \tag{1}$$

$$w_j = m_j / \sum_{j=1}^{N} m_j, \tag{2}$$

where

$\bar{x}_i$ is the weighted mean of the content of the i-th component in the mixture;
x_{ij} is the content of the i-th component in the j-th component in the mixture;
w_j is weight coefficients;
m_j is the mass of the j-th component in the mixture.

ISO Guide 35: 2017 also offers a formula for estimating the standard uncertainty due to characterization using the gravimetric preparation procedure of a binary mixture [7, (13)], for the case when $N = 2$.

However, within the framework of the GUM methodology [8], the work [9] shows that the square of the combined standard uncertainty of the output value x_j can be represented for an unlimited number of components ($N < \infty$) in general:

$$u_{\bar{x}_i}^2 = \sum_{j=1}^{N} \left[m_j^2 \cdot u_{x_{ij}}^2 + \left(x_{ij} - \bar{x}_i\right)^2 \cdot u_{m_j}^2 \right] \Big/ \left(\sum_{j=1}^{N} m_j \right)^2, \tag{3}$$

where

$u_{x_{ij}}$ is the standard uncertainty of the content of the j-th component in the mixture;
u_{m_j} is the standard measurement uncertainty of the mass of the j-th component in the mixture.

3 Materials and Methods

In order to test the proposed approach, a multicomponent RM was prepared, consisting of a mix of solutions of chemical elements (V, Cr, Mn, Fe, Co, Ni, Cu, Zn, Cd) in a matrix of 5% nitric acid (*blank*). Mono-element solutions were gravimetrically prepared from pure metals corrected for the air buoyancy in weighing the metal sample intake and final solution. For the preparation of solutions, containers of high density polyethylene were used, purified using non-boiling distillation of nitric and hydrofluoric acids and deionized water. The mass fraction of the main component in the basic material was determined according to the scheme "100% minus combined impurities" using the results of measurements by the mass spectrometry method with inductively coupled plasma (ICP MS, PerkinElmer Inc. NexION 300D). For the determination of gas-forming impurities (N, O, H), the reducing melting method (CGHE, Eltra GmbH ELEMENTRAC ONH-p) was used. The evaluation of this value and associated uncertainty was tested by the following formulas:

$$w = 100 - \sum_{i=1}^{n} z_i - \sum_{j=1}^{r} \frac{y_j}{2}, \tag{4}$$

$$U(w) = 2 \cdot u_c(w) = 2 \cdot \sqrt{\sum\nolimits_{i=1}^{n} u(z_i)^2 + \sum\nolimits_{j=1}^{r} u(y_j)^2}, \tag{5}$$

where

$u_c(w)$ is the combined standard uncertainty of the content of the main component in the starting material;
z_i is the mass fraction of the *i-th* detected impurity;
n is the number of detected impurities;
y_j is the detection limit established for the *j*-th impurity according to the 3σ criterion;
r is the number of detection limits;
$u(z_i)$ is the standard measurement uncertainty of the *j*-th impurity;
$u(y_j) = y_j/2$ is the standard uncertainty caused by the establishment of a detection limit in the *j*-th impurity.

When $n + r = 91$, the most complete assessment of purity in the characterization of comparison standards is achieved.

The mass fractions of the target impurity components in each solution were determined in the same way using ICP MS and the mass fraction of the major component according to the formula:

$$x = K_{stab} \cdot K_{homo} \cdot K_{evap} \cdot \left(w_0 + \frac{m_n \cdot w \cdot \left(\frac{\rho_a}{\rho_r} - 1 \right)}{m_r \cdot \left(\frac{\rho_a}{\rho_n} - 1 \right)} \right), \tag{6}$$

where

K_{stab}, K_{homo}, K_{evap} are coefficients associated with temporal stability, homogeneity and possible evaporation of the solution during the preparation process, respectively (in the calculations they are taken to be equal to 1, and their uncertainties are determined experimentally);
w_0 is the mass fraction of the main component in the form (in the example below, the value is taken to be equal to 0 with the uncertainty expressed in the form of the detection limit);
m_n is the mass of the starting metal sample intake (pure metal);
m_r is the mass of the solution;
w is the mass fraction of the main component in the starting material;
ρ_a, ρ_r, ρ_n are the densities of air, solution and starting material, respectively.

The characteristic budget of uncertainty for the analyte content in such a solution is assembled in Table 1. It can be seen that the relative expanded uncertainty of the element mass fraction in such a solution is lower than 0.05%.

The original data of the mixture preparation are presented in Table 2.

Table 1 Uncertainty budget of the *primary calibration solution* preparation

Source		Value	u	Unit	c	u × c	u × c (%)
Description	Symbol						
Mass fraction in starting material	w	999.940	0.060	mg/g	1.0×10^{-3}	6.0×10^{-5}	12.7
Sample mass	m_n	1.0019	0.0002	g	1.0	2.0×10^{-4}	42.1
Solution mass	m_r	1,000.901	0.002	g	-1.0×10^{-3}	1.5×10^{-6}	0.3
Air density	ρ_a	1.1336	0.0079	kg/m^3	-8.7×10^{-4}	6.9×10^{-6}	1.5
Starting material density	ρ_n	8,920	500	kg/m^3	-1.4×10^{-8}	7.1×10^{-6}	1.5
Solution density	ρ_r	1,017	5	kg/m^3	1.1×10^{-6}	5.5×10^{-6}	1.2
Mass fraction in blank	w_0	0	5.0×10^{-5}	mg/g	1.0	5.0×10^{-5}	10.5
Possible solution Inhomogeneity	K_{homo}	1	1.0×10^{-5}	mg/g	1.0	1.0×10^{-5}	2.1
Solution stability*	K_{stab}	1	3.3×10^{-5}	mg/g	1.0	3.3×10^{-5}	7.0
Possible evaporation	K_{evap}	1	1.0×10^{-4}	mg/g	1.0	1.0×10^{-4}	21.1
Mass fraction in solution	x	0.99995		mg/g	* from evaporation weight loss measurement Nalgene 125 PEHD bottles for 1 month		
Combined standard uncertainty	u	0.00024		mg/g			
Expanded uncertainty (k = 2, P = 0.95)	U	0.00048		mg/g			

For the estimation of the uncertainty of the mass fraction of the target components in the mixture by formula (3), the standard mass measurement uncertainty was taken to be 0.0005 g, and the uncertainties of the mass fractions of the components listed in Table 3 were used. The values of the standard uncertainties of the main component mass fractions in each of the solutions are taken to be equal to 0.1%, since such tasks commonly imply the use of the so-called *secondary calibration solutions*, rather than reference solutions.

4 Results

The proposed approach was validated using the Monte Carlo simulation (MMC) [10] for 1×10^6 values of the content of each element having a normal distribution of input

Table 2 Mixture preparation data

Mass fraction in solution (ppm)										
Element	*Primary calibration solution*									Blank
	V	Cr	Mn	Fe	Co	Ni	Cu	Zn	Cd	
V	**1,001**	0.071	0.070	0.070	0.070	0.070	0.070	0.070	0.070	0.070
Cr	0.14	**1,000**	0.13	0.094	0.09	0.09	0.09	0.09	0.09	0.090
Mn	0.045	0.045	**1,000**	0.057	0.046	0.045	0.045	0.05	0.045	0.045
Fe	0.112	0.111	0.167	**999**	0.108	0.218	0.102	0.30	0.102	0.102
Co	0.021	0.021	0.0230	0.031	**1,000**	0.024	0.021	0.021	0.021	0.021
Ni	0.135	0.1360	0.1370	0.141	0.175	**1,000**	0.135	0.138	0.135	0.135
Cu	0.062	0.062	0.211	0.0631	0.066	0.062	**1,001**	0.046	0.0010	0.062
Zn	0.005	0.0020	0.109	0.035	0.0010	0.005	0.004	**1,000**	0.1120	0.111
Cd	0.019	0.019	0.019	0.019	0.019	0.019	0.019	0.199	**1,000**	0.019
Solution mass (g)	9.9894	9.9998	10.0000	10.0011	10.0014	9.9938	9.9898	10.0004	10.0006	10.0754

Table 3 Information on the uncertainty of input data

Combined standard uncertainty of the mass fraction in solution (ppm)										
Element	*Primary calibration solution*									Blank
	V	Cr	Mn	Fe	Co	Ni	Cu	Zn	Cd	
V	1	0.000	0.006	0.006	0.006	0.006	0.006	0.006	0.006	0.006
Cr	0.01	1	0.016	0.028	0.0045	0.0045	0.0045	0.0045	0.0045	0.005
Mn	0.0036	0.0036	1	0.011	0.0092	0.0036	0.0036	0.009	0.0036	0.004
Fe	0.0056	0.00555	0.008	1	0.016	0.013	0.0051	0.01	0.0051	0.005
Co	0.00168	0.00168	0.0046	0.005	1	0.004	0.00168	0.00168	0.00168	0.002
Ni	0.00675	0.0082	0.00685	0.00703	0.00875	1	0.00675	0.0069	0.00675	0.007
Cu	0.00496	0.00496	0.013	0.0050	0.006	0.00496	1	0.008	0.0002	0.005
Zn	0.00555	0.0003	0.007	0.010	0.0003	0.00555	0.00555	1	0.0067	0.006
Cd	0.00152	0.00152	0.00152	0.00152	0.00152	0.00152	0.00152	0.016	1	0.002

values, with the standard deviation being numerically equal to their total standard uncertainty. The output value has a distribution close to normal with a coverage factor of about 1.96, depending on the input data (Fig. 1).

Table 4 displays the results of calculations using formulas (1–3), the averaged data obtained by MMC and the relative deviation of these assessments.

The comparison showed the compatibility of the obtained estimates and, consequently, the correctness of the weighted means of the components in the mixture and calculation of their standard uncertainty according to (1–3). At the same time,

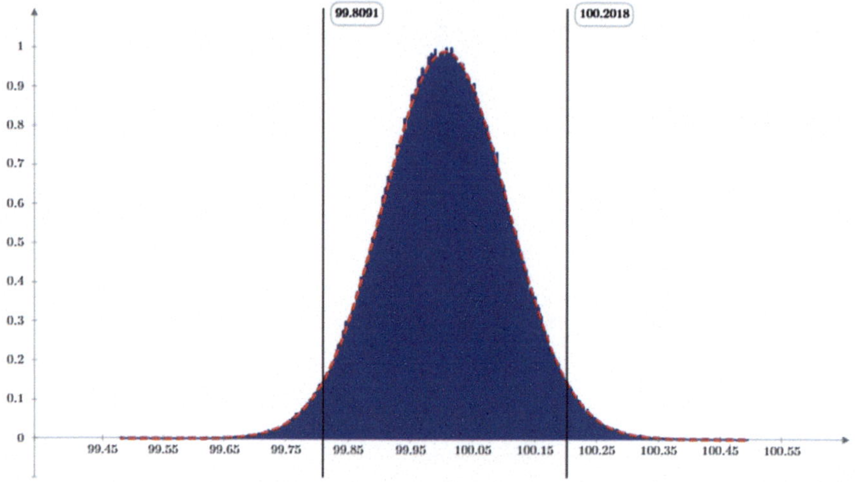

Fig. 1 Distribution of the output value in the MMC simulation

Table 4 Processing results

Element	Formula (1–3) evaluation		MMC evaluation		Relative deviation of results	
	W (ppm)	U (ppm)	W (ppm)	U (ppm)	δW (%)	δU (%)
V	100.0054547	0.199945401	100.0055416	0.200302	-8.6885×10^{-5}	−0.17803086
Cr	100.0369303	0.200243396	100.036775	0.2007786	0.000155208	−0.26656413
Mn	99.99020344	0.200158038	99.9901002	0.2008064	0.000103252	−0.32287893
Fe	99.99147791	0.200214825	99.9915292	0.2006416	-5.1296×10^{-5}	−0.21270494
Co	99.98240354	0.200156003	99.9822666	0.200768	0.000136962	−0.304828
Ni	100.0131538	0.200044727	100.0130788	0.2007424	7.50212×10^{-5}	−0.34754619
Cu	100.0098578	0.199956752	100.0098408	0.200462	1.69785×10^{-5}	−0.25204201
Zn	99.99087815	0.200161564	99.9908142	0.2007514	6.3952×10^{-5}	−0.29381416
Cd	99.9895358	0.200160763	99.9895264	0.2005146	9.40462×10^{-6}	−0.17646446

it is shown that the most significant contribution to the uncertainty of certified values is made by the uncertainty associated with the mass fractions of the certified components in their initial solutions (the relative contribution is more than 90%). Thus, the uncertainty of the certified values is composed of uncertainties associated with the gravimetric preparation of pure substance solutions calculated by the "100% minus combined impurities" scheme and those induced by the stability of these solutions and the results of comparative measurements using high-performance inductively coupled plasma optical emission spectroscopy (ICP OES, PerkinElmer Inc. OPTIMA 7300DV).

Then, if the mass of mixture components is determined sufficiently accurately, in order to avoid loss of precision of the final results, the sum of the impurity of the target element in the components of the mixture should not significantly exceed the value of the standard uncertainty of the target component content in the main solution. In this case, the accuracy of determining the composition of the impurity is not significant. Conversely, when the minimum accumulation of the target component impurities is impossible to achieve, the uncertainty of the combined content of impurities must not exceed ½ of the value of the standard uncertainty of the content of the target component in its main solution.

Estimates of components received by the preparation procedure are confirmed by measurements of these quantities by the ICP OES method (x_l, $L = 2$) (Fig. 2).

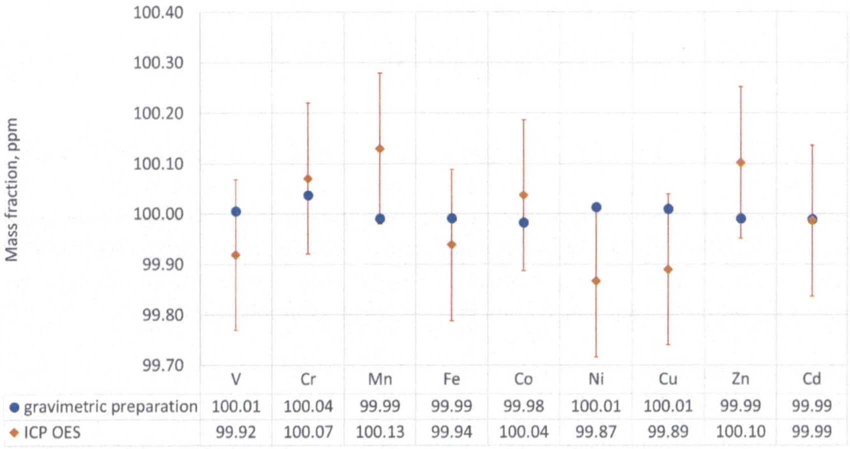

Fig. 2 Comparison of calculated data and ICP OES results

5 Discussion and Conclusions

It is worth noting that the question remains what to consider as the RM certified value: (1) the results obtained according to the preparation procedure; (2) the results of ICP OES measurements; (3) their mean or weighted mean.

There exists a widespread opinion that the uncertainty of the certified RM value cannot be better than the uncertainty associated with the measurement of this quantity, which can be demonstrated by the producer of that RM. We believe the approach, according to which the results of direct measurements, rather than the calculated data, are taken into account in calculations, to be rather unreasonable. The thing is that any additional knowledge about the measured (certified) value can bring the researcher closer to more reliable estimates. Therefore, such information should never be wasted.

It seems sensible to consider the final result to be the weighted mean of the abovementioned estimates (x_r) with weighting coefficients associated with their standard uncertainties (u_l) and the square of the combined standard uncertainty of the quantity x_r as a component of the uncertainty of the certified RM value due to characterization:

$$x_r = \sum_{l=1}^{L} \frac{x_l}{u_l^2} \Bigg/ \sum_{l=1}^{L} \frac{1}{u_l^2}, \tag{7}$$

$$u_{x_r}^2 = 1 \Bigg/ \sum_{l=1}^{L} \frac{1}{u_l^2}. \tag{8}$$

This approach is generally similar to that used in the characterization of mono-element RMs [11, 12], where the certified value is obtained as the weighted mean of the results of gravimetric preparation of a solution from a high-purity compound (with the established content of the main component) and those obtained using ICP OES (the spectrometer being calibrated by four primary standards from a high-purity compound with the established content of the main component).

Acknowledgements We are grateful to Tatiana N. Tabatchikova, the leading engineer of UNIIM, for providing measurement results obtained using the ICP MS method.

Author Contribution Statement Migal P. V.: Identification of the paper concept and metrology, data analysis, collection of literature data, obtainment of experimental data, preparation of the manuscript draft, computer manuscript preparation, preparation of the final manuscript.

Medvedevskikh S. V.: Concept and initiation of the research, methodological support, critical analysis and text processing.

Firsanov V. A.: Methodical and mathematical support for the works described in the article.

Conflict of Interest One of the authors is a member of the Editorial Board of these proceedings.

References

1. Database “Reference standards in the form of high-purity substances”. Ural Research Institute for Metrology, Ekaterinburg. https://www.uniim.ru/db_etalons_ref
2. Medvedevskih SV, Sobina EP, Migal PV, Goryaeva LI, Gorbunova EM, Tabatchikova TN, Sobina AV, Firsanov VA, Medvedevskikh MYu, Krasheninina MP (2014) On the use of pure inorganic substances in metrology of analytical measurements. Ref Mater 3:58–67 (in Russ)
3. Matschat R, Czerwensky M, Pattberg S, Heinrich HJ, Tutschku S (2002) High purity metals as primary calibration materials for elemental analysis—their importance and their certification. Mater Trans 43(2):90–97. https://doi.org/10.2320/matertrans.43.90
4. Kipphardt H, Matschat R, Vogl J, Gusarova T, Czerwensky M, Heinrich HJ, Hioki A, Konopelko LA, Methven B, Miura T et al (2010) Purity determination as needed for the realization of primary standards for elemental determination—status of international comparability. Accred Qual Assur 15(1):29–37. https://doi.org/10.1007/s00769-009-0557-0
5. Vogl J et al (2018) Establishing comparability and compatibility in the purity assessment of high purity zinc as demonstrated by the CCQM-P149 intercomparison. Metrologia 211–221. https://doi.org/10.1088/1681-7575/aaa677
6. Salit ML, Turk GC, Lindstrom AP, Butler TA, Beck CM II, Norman B (2001) Single-element solution comparisons with a high-performance inductively coupled plasma optical emission spectrometric method. Anal Chem 73:4821–4829. https://doi.org/10.1021/ac0155097
7. ISO Guide 35:2017 Reference materials—guidance for characterization and assessment of homogeneity and stability
8. JCGM 100:2008 Evaluation of measurement data—guide to the expression of uncertainty in measurement
9. Medvedevskih SV, Firsanov VA (2018) Determination of the integral content of the components of substances in the control, recording, and monitoring of production and generation of waste products in industrial quantities. Meas Tech 61:182–191. https://doi.org/10.1007/s11018-018-1407-4
10. JCGM 101:2008. Evaluation of measurement data—supplement 1 to the “Guide to the expression of uncertainty in measurement”—propagation of distributions using a Monte Carlo method
11. Standard reference materials. Spectrometry, single element standard solutions: NIST. https://www.nist.gov/srmors/viewTable.cfm?tableid=39
12. NIST (2000) NIST Special Publication 260-136. https://www.nist.gov/sites/default/files/documents/srm/SP260-136.PDF

Use of Reference Materials

Bismuth-Containing Glassy Reference Materials for Establishing the Content of Transition Elements Using Luminescent-Based Methods

Valentina V. Borisova, Valentin A. Kutvitsky, Elena V. Mironova, Inna A. Romanova and Olga I. Ostanina

Abstract Modern express methods of analytical control are widely used for monitoring the quality of oxide materials. However, the possibilities of these methods are currently limited by the lack of versatile homogeneous reference materials (RMs). Such RMs can most effectively be produced by vitrification methods. In this connection, glassy RMs based on bismuth-borate systems are of great interest. In earlier research, bismuth-borate RMs were used for comparison purposes in determining the composition of complex oxide systems using the X-ray fluorescence method. In this work, we consider issues associated with the synthesis and application of bismuth-borate glassy RMs for determining the content of transition elements in oxide compounds and mixtures using luminescence analysis. Luminescence analysis was used as a research method due to its high selectivity and low detection limits with respect to elemental impurities in controlled objects (usually constituting 1×10^{-6}–1×10^{-4} wt%). Bismuth-borate glassy RMs are shown to be capable of luminescence under UV radiation in the visible range at a temperature of 77 K. It is established that the introduction of the oxides of transition and rare earth elements (REE) into the RM composition leads to the quenching of the exciton emission band in the studied glasses at 77 K, as well as to the appearance of bands typical of REE ion radiation both at 77 and at 298 K. Glassy RMs doped with transition-element oxides are found to quench the luminescence of both bismuth and REE ions at a temperature of 77 K. On the basis of the results obtained, measurement procedures for determining the content of quenching elements in oxide materials have been developed. The metrological characteristics of these procedures have been evaluated.

Keywords Bismuth-borate glassy reference materials · Luminescence · Luminescence spectra · Rare-earth ions · Luminescence quenching · Transition elements · Calibration dependency · Metrological characteristics

V. V. Borisova (✉) · V. A. Kutvitsky · E. V. Mironova · I. A. Romanova · O. I. Ostanina
MIREA – Russian Technological University (RTU MIREA), Moscow, Russia
e-mail: vv_borisova@mail.ru

S. Medvedevskikh et al. (eds.), *Reference Materials in Measurement and Technology*,
https://doi.org/10.1007/978-3-030-32534-3_18

Abbreviations

XRF X-ray fluorescence
REE Rare earth element
AA Atomic absorption
RM Reference material
REI Rare-earth ion
HTS High-temperature superconductor
LA Luminescence analysis

1 Introduction

Intensive development in the sphere of manufacturing new materials for modern technology requires improved methods for monitoring their quality and, above all, effective compositional reference materials (RMs) for instrumental analysis methods.

A promising approach to the synthesis of compositional RMs consists in preparing them via vitrification from the molten state using a mixture of bismuth and boron oxides. RMs obtained on the basis of such glass-forming matrices have been successfully used in the X-ray fluorescence (XRF) analysis of elements in the composition of complex oxide materials [1].

Bismuth-borate glasses are characterised by the phenomenon of luminescence at cryogenic temperatures. Changes in the glass-intrinsic luminescence spectrum following the introduction of transition metal impurities allow the quantity of these impurities to be determined.

2 Materials and Methods

In order to obtain a glassy matrix, high-purity bismuth and boron oxides were thoroughly mixed in an agate mortar together with other oxide mixture components and subsequently transferred to a platinum crucible. The crucible was then placed into a muffle furnace, where it was subjected to stepwise heating up to a melting temperature (T_m = 1273–1373 K) and held at this temperature for an hour. Afterwards, the melt was poured into a crucible and, following the glass transition process, placed in a quenching furnace, where post-quenching annealing was performed at a temperature of 673 K for 12 h, followed by cooling in an oven to room temperature. According to the described technology, RMs containing 70 wt% Bi_2O_3, (30-X-Y) wt% B_2O_3, X (0–2%) wt% Ln_2O_3 (Ln = Sm, Eu, Tb, Dy, Tm), Y (0–1%) wt% (Y = CuO, Fe_2O_3, Co_2O_3) were obtained. The quality of the obtained comparison RMs was assessed according to the results of X-ray Phase Analysis (DRON-3M diffractometer, Burevestnik, Russia) and by measuring the microhardness value (PMT-3M microhardness tester, LOMO, Russia). The homogeneity of the glassy RMs was estimated according to the algorithm established in GOST 8.531 [2] based on the results of microhardness measurements using the Vickers method. In order to determine the content of controlled elements in oxide materials, the luminescent analysis method was used.

The measurements were carried out using a system, in which an SVD-120A Hg-lamp placed in a lightproof case served as an excitation source. In order to isolate the ultraviolet portion of the spectrum (transmission limit: 250–400 nm), a UFS-2 wideband glass optical filter (Arli, Russia) was placed at the source emission point after the aperture. The transformation of the excitation flow was carried out using quartz lenses (KU 1 lens material) and a mirror. As a result, a reduced image of the source aperture (about 10 mm) was obtained on the material surface. Then, the excitation flow passed through an optical modulator, which performed its 100% modulation and produced sync pulses to start the master oscillator. The luminescent radiation was collected using a lightguide and was refocused onto the slit of a SPM-2 prism monochromator (Carl Zeiss Jena GmbH, Germany). The monochromator was supplemented with a spectrum scanning system and a photoelectric recording unit. In order to increase the sensitivity of this system, a phase-sensitive amplification method with synchronous detection was used. The recording of the resulting spectra was carried out using a KSP-4 self-recording potentiometer (Manometer, Russia). The design of the cell holder allowed luminescent studies to be carried out both when cooling the cell with liquid nitrogen and at room temperature.

X-ray fluorescence (XRF) and atomic absorption (AA) analyses were used as alternative methods for determining the content of transition elements in the materials under study (in particular, copper). XRF analysis was performed using a VRA-33 X-ray spectrometer (Carl Zeiss Jena GmbH, Germany). Here, primary radiation was formed in an X-ray tube having a chrome or copper anode, with the heating voltage and current of the X-ray tube being 40 kV and 35 mA, respectively. Scintillation and proportional flow counters used in tandem served as detectors. LiF or graphite crystals were used for crystal analysis.

For the AA analysis, a Kvant-Z.ETA atomic absorption spectrometer with thermal activation of the sample in a Lvov cell in argon atmosphere (KORTEK, Russia) was used.

Resistance to environmental effects was evaluated according to R 50.2.031 [3]. The luminescent radiation intensity of the glassy RM was used as a certified characteristic. The value of the certified RM characteristic varied within the RM error limits provided that the storage and use conditions were ensured during the 3-year period.

3 Results and Discussion

To the best of our knowledge, no luminescent studies on the possibility of using bismuth-borate glassy RMs for determining transition elements and rare earth ions

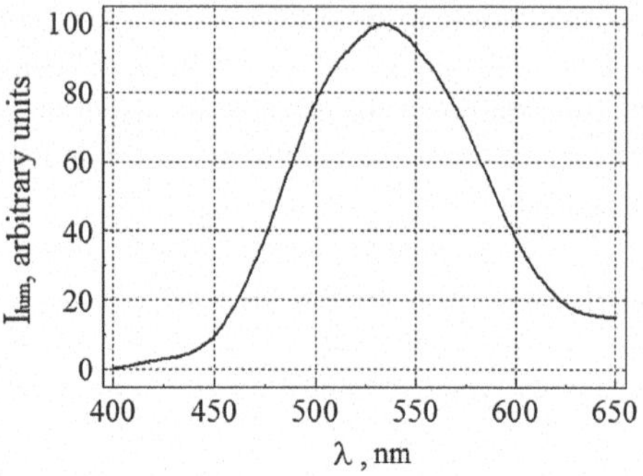

Fig. 1 Luminescence spectrum of a bismuth-borate glassy RM at T = 77 K

(REI) in complex oxide systems using the quenching effect of the intrinsic luminescence band of REIs and the exciton emission band of Bi^{3+} ion radiation at 77 K have been carried out thus far.

At T = 77 K, the luminescence spectrum of a glassy RM for the composition of 70 wt% Bi_2O_3–30 wt% B_2O_3, which is formed when the material is irradiated by light having a wavelength corresponding to the absorption band edge, constitutes a wide asymmetric band having a maximum of 520 nm in the 400–600 nm range (Fig. 1). The width of the luminescence band is associated with electron-phonon interaction processes, with the shape of the luminescence spectrum depending on the Huang-Rhys parameter [4].

In terms of its shape and maximum, the luminescence spectrum of bismuth-borate glassy RMs is identical to that of crystals having a eulytine structure. This is explained by the fact that the Bi^{3+} ion is a mercury-like cation with a $5d^{10}6s^2$ electron configuration, with the energy scheme of Bi^{3+} levels corresponding to the general scheme of other mercury-like ions (Tl^+, Pb^{2+}) [5]. The ground state of the bismuth ion is $^1S_0$. The excitation levels correspond to the 6s6p configuration: triplet $^3P_0$, $^3P_1$, $^3P_2$ and singlet $^1P_1$. The $^1S_0 \rightarrow {}^3P_1$ and $^1S_0 \rightarrow {}^1P_1$ transitions are realised in the optical absorption and excitation bands of photoluminescence, with the former taking place at longer wavelengths. A characteristic property of bismuth-containing crystals is a wideband luminescence associated with the lowest permissible transition $^3P_1 \rightarrow {}^1S_0$. Therefore, the radiation process in bismuth-borate glassy RMs can be attributed to the $^3P_1 \rightarrow {}^1S_0$ transition of the Bi^{3+}ion. It can be assumed that such glassy RMs contain the $[BiO_6]^{9-}$ type clusters similar those in germanoeulytine, which participate in the radiation process. It can be concluded, therefore, that the glassy RMs under study feature an exciton charge-transfer mechanism.

Since the luminescent properties of bismuth-borate glasses have been successfully used for the phase analysis of bismuth-containing systems, they can be used in elemental analysis for the analytical determination of luminescence quenching elements that suppress the emission of activators. The luminescence quenchers suppress the emission of activators due to the presence of a charge-transfer band, whose maxima are in the UV region of the spectrum; however, their intensity is so great that the tail of these bands extends into the visible and even into the near infrared region

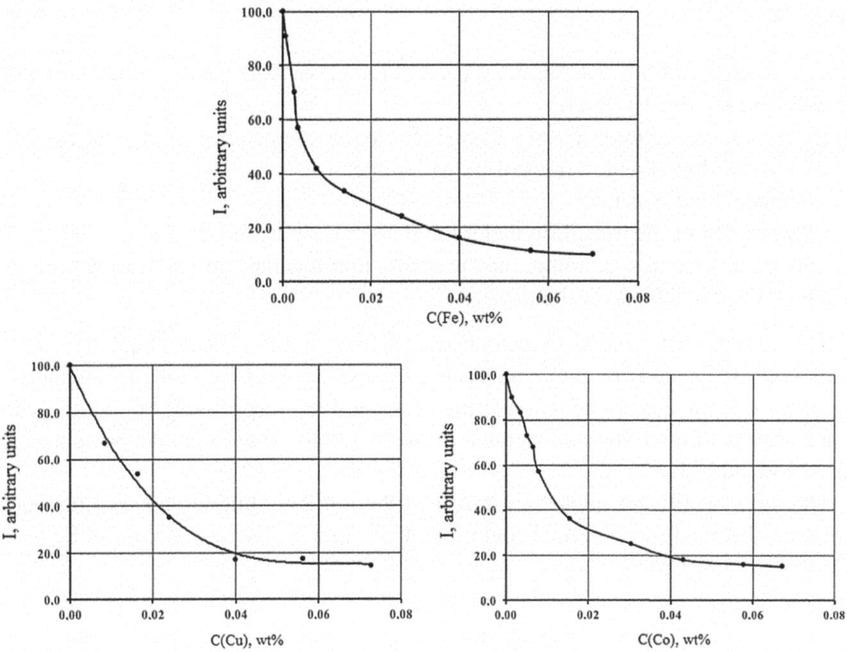

Fig. 2 Dependence of the luminescence intensity of bismuth-borate glassy RMs doped with Fe (III), Co (III), Cu (II) oxides

(Fe^{3+}). The charge-transfer band quenches the luminescence of all activators whose emission bands overlap with it [6]. The quenching of a particular type is usually caused by ions of variable oxidation state, i.e. transition metal ions, specifically Fe, Co, Ni.

Upon the introduction of the individual oxides of transition elements into the bismuth-borate matrix (CuO, Fe_2O_3, Co_2O_3), the effect of quenching the intensity of the exciton luminescence band Bi^{3+} ($\lambda_{max} = 520$ nm) was established. The dependencies of the luminescence intensity of bismuth-borate glassy RMs doped with oxides of some transition elements are shown in Fig. 2.

It is established that the dependence of the relative luminescence intensity of Bi^{3+} ions on the concentration of the transition element in bismuth-borate glassy RMs has a complex character. The analysis of the experimental results indicates that a decrease in the radiation intensity of Bi^{3+} ions occurs according to the hyperbolic law; therefore, a Stern-Volmer equation is proposed for describing the dependence of the relative luminescence on the content of the transition element ion:

$$\frac{I_0}{I} = 1 + a \cdot C^b, \tag{1}$$

where:

I_o is the intensity of intrinsic luminescence of the glassy RM having composition of 70 wt% Bi_2O_3–30 wt% B_2O_3;
I is the luminescence intensity of glassy RMs having composition of 70 wt% Bi_2O_3–30 wt% B_2O_3 doped with oxides of transition metals;
a is the quenching constant;
C is the content of the transition metal ion in the glassy material;
b is the parameter that accounts for the interaction mechanism (indicating various means of the quenching mechanism).

The calibration dependences are constructed using the coordinates $\ln[[I_o/I) - 1] = f(\ln C)$ (where I_o and I are the luminescence intensities of the Bi^{3+} ion in the material that does not contain/contains transition element ions, respectively; *C* is the ionic concentration of the transition element in the material). These data are presented in Fig. 3 and Table 1.

The most significant quenching effect of the intrinsic luminescence of bismuth ions was observed for materials containing Cu^{2+} ions, in the series Cu–Co–Fe by a factor of around 1.5.

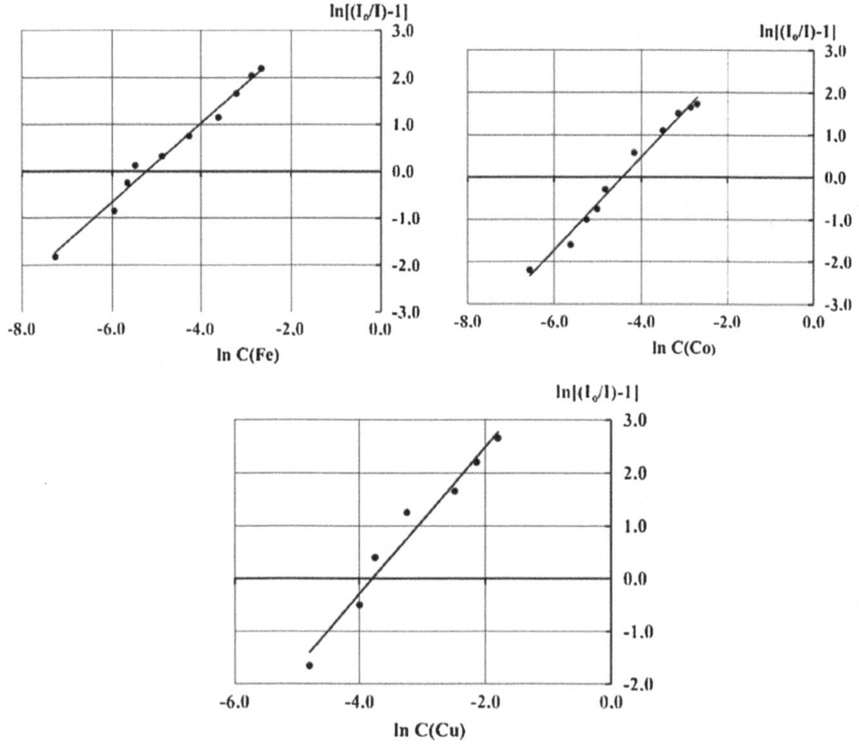

Fig. 3 Calibration dependencies for the luminescence determination of Fe, Co, Cu

Table 1 Calibration dependence equations for luminescence quenching of Bi^{3+} ions in bismuth-borate glasses doped with transition element ions at T = 77 K (P = 0.95, n = 5)

Object of analysis	Analysed component	Calibration dependence equation in the form $\ln[(I_o/I) - 1] = a + b \cdot [\ln C]$	R^2	C_{min} (wt%)
Fe_2O_3	Fe	$\ln[I_o/I - 1] = 4.41 + 0.85 \cdot \ln C$ (Fe)	0.97	0.001
Co_2O_3	Co	$\ln[I_o/I - 1] = 4.85 + 1.09 \cdot \ln C$ (Co)	0.99	0.001
CuO	Cu	$\ln[I_o/I - 1] = 5.22 + 1.25 \cdot \ln C$ (Cu)	0.97	0.003

On the basis of the established luminescence quenching effect of bismuth ions by copper ions in bismuth-borate glassy materials, a measurement procedure was developed for the low-temperature luminescence determination of the copper content in bismuth-containing high-temperature superconductors (Bi-HTS). Bi-HTS are of great practical importance due to their sufficient plasticity, as well as the absence of highly volatile (Hg-, Tl-HTS), toxic (Hg-, Tl-HTS) and expensive elements (Y-, REE-HTS). HTS constitute phases of variable composition. Their real chemical composition differs from that reflected by their chemical formula, with the bismuth, strontium and calcium atomic contents being, as a rule, mutually dependent [7]. Glassy materials doped with CuO and Bi-HTS having composition of $Bi_{2+x+z}Sr_{2-x-y}Ca_{n-1-z+y}Cu_{n-y}O_{4+2n+\delta}$ were prepared and their luminescent properties studied. On the basis of the obtained experimental data, the calibration dependence equations presented in Table 2 were constructed.

The confidence interval boundaries for coefficients *a* and *b* were evaluated at a given significance level $\beta = 0.01$ using the Laplace function. In this case, the natural logarithm of coefficient *a* can be $3.65 < \ln a < 5.12$, while that of coefficient *b* is $1.19 < \ln b < 1.47$. A comparison of the equations describing these calibration dependencies showed that they overlapped within the confidence interval. Consequently, RMs doped only with CuO can be used for determining copper in the composition of Bi-HTS, which contain, along with copper, elements that are not involved in quenching Bi^{3+} ions.

Table 2 Equations for the calibration dependencies of the luminescence of bismuth-borate glasses at T = 77 K on the copper content (n = 5, P = 0.95)

Object of analysis	Calibration dependency the equation in the form $\ln[(I_o/I) - 1] = a + b \cdot [\ln C]$	R^2	C_{min} (wt%)
CuO	$\ln[I_o/I - 1] = 4.88 + 1.25 \cdot \ln C(Cu)$	0.997	0.001
$Bi_2Sr_2CuO_6$	$\ln[I_o/I - 1] = 4.91 + 1.33 \cdot \ln C(Cu)$	0.996	0.001
$Bi_2Sr_2CaCu_2O_8$	$\ln[I_o/I - 1] = 4.12 + 1.31 \cdot \ln C(Cu)$	0.995	0.003
$Bi_2Sr_2Ca_2Cu_3O_{10}$	$\ln[I_o/I - 1] = 4.25 + 1.34 \cdot \ln C(Cu)$	0.996	0.001

Table 3 Metrological characteristics of the determination of copper content in $Bi_2Sr_2CuO_6$, using various instrumental methods of analysis (P = 0.95, n = 5)

Method of analysis	Accepted reference value C(Cu) (wt%)	Arithmetic mean of analysis results (Cu) (wt%)	Estimation of standard deviation of repeatability σ_r (wt%)	Estimation of the standard deviation of interlaboratory precision σ_r (wt%)	Trueness estimation $\pm\Delta_c$ (wt%)	Accuracy estimation $\pm\Delta$ (wt%)
LA	0.13	0.14	0.0024	0.0024	0.0053	0.0071
	0.17	0.17	0.0029	0.0029	0.0065	0.0086
	0.28	0.27	0.0041	0.0041	0.0094	0.012
	0.34	0.34	0.0051	0.0051	0.012	0.015
XRF analysis	0.13	0.14	0.0014	0.0014	0.0035	0.0045
	0.17	0.16	0.0016	0.0016	0.0043	0.0053
	0.28	0.29	0.0029	0.0029	0.0074	0.0094
	0.34	0.33	0.0033	0.0033	0.0086	0.011
AA analysis	0.13	0.14	0.0028	0.0028	0.0059	0.0080
	0.17	0.16	0.0032	0.0032	0.0069	0.0093
	0.28	0.29	0.0058	0.0058	0.012	0.017
	0.34	0.33	0.0066	0.0066	0.014	0.019

Since the power exponent in Eq. (1) is greater than one, it can be assumed that copper ions simultaneously participate in quenching processes by various mechanisms (migration and multipole). Since the value of *b* is slightly greater than one, the contribution to quenching due to multipole interaction can be disregarded [7].

The metrological characteristics of the afore-described procedure for determining copper in Bi-HTS having the composition of $Bi_2Sr_2CuO_6$ using a low-temperature luminescence method (Table 3) were evaluated according to RMG 61-2010 [8]. The error of the luminescence determination of copper in HTS does not exceed 0.05 in relative units, with a detection limit being 0.001 wt%.

The trueness of the measurement procedure for the low-temperature luminescence determination of copper was additionally established by comparing the calculated results with those obtained using the X-ray fluorescence and atomic absorption methods (Table 3).

The emergence of new high-temperature superconducting phases in the Sr–Ln–Cu–O and Hg–Ce–Sr–Ca–Ln–Cu (Ln = La–Lu) systems necessitates the monitoring of constituent elements. It was established that the introduction of REI oxides (Sm, Eu, Tb, Dy, Tm) into the bismuth-borate glassy matrix leads to a decrease in the intensity of the luminescent radiation of Bi^{3+} ions with an increase in the REI content at 77 K (Fig. 4a), which conforms to the Stern-Volmer law (as is the case when transition elements are present in glassy materials). The Dy^{3+} ions are shown to have the smallest influence on the luminescence of Bi^{3+} ions. At room temperature, two narrow emission bands having a maximum at 485 nm, corresponding to the $^4F_{9/2} \rightarrow {}^6H_{15/2}$ transition, were observed, as well as a maximum of 580 nm, corresponding to the $^4F_{9/2} \rightarrow {}^6H_{13/2}$ transition (Fig. 4b).

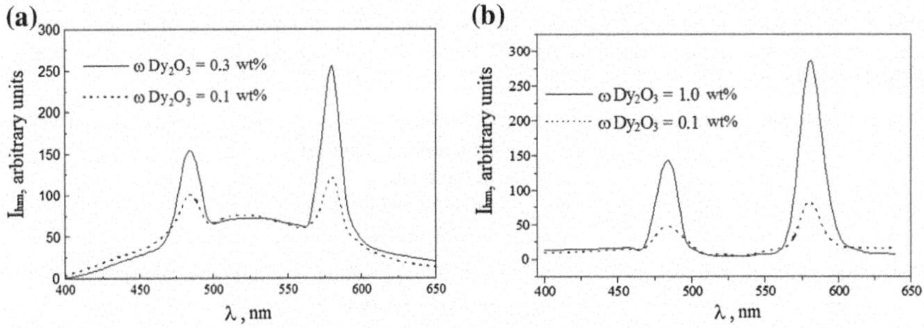

Fig. 4 Luminescence spectra of bismuth-borate glasses doped with Dy_2O_3. **a** At a temperature of 77 K, **b** at a temperature of 298 K

In order to study the joint effect of the REI and the transition element ion on the intrinsic luminescence of glassy RMs, bismuth-borate materials were prepared with the addition of 0.5 wt% Dy_2O_3 and CuO in the concentration of 0.001–0.25 wt% into the material matrix. The Dy_2O_3 content, equal to 0.5 wt%, corresponded to its maximum possible content in the range of the linear dependence of the intrinsic luminescence of Dy^{3+} on its concentration at room temperature. The luminescent properties of the obtained glassy materials were studied at temperatures of 77 and 298 K (Fig. 5).

It is established that a 90% decrease in the luminescence intensity at a temperature of 298 K is achieved when the Dy^{3+} content of the materials equals 0.06 wt%. The calibration dependence presented in Table 4 was constructed using logarithmic coordinates $\ln[[I_o/I) - 1] = f(\ln C)$ in order to determine the copper content in the 0.001–0.06 wt% range (where I_o is the luminescence intensity of Dy^{3+} in a material that does not contain Cu^{2+}; I is the luminescence intensity of Dy^{3+} in a material that contains Cu^{2+}, C is the concentration of Cu^{2+} ions).

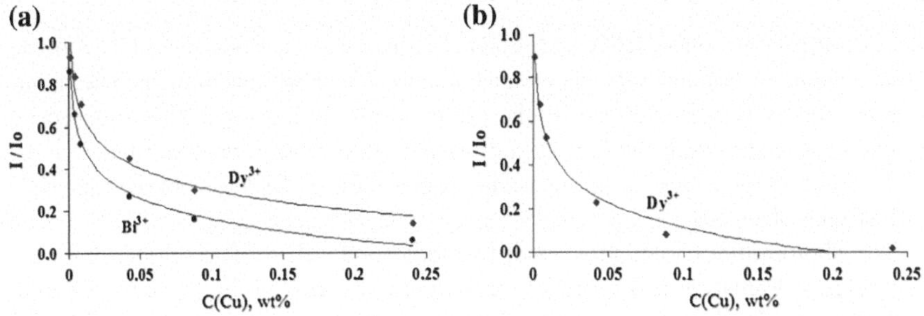

Fig. 5 Dependencies of the luminescence intensity of Bi^{3+} and Dy^{3+} in bismuth-borate glassy RMs on the copper content. **a** At a temperature of 77 K, **b** at a temperature of 298 K

Table 4 Calibration dependence equation for the luminescent determination of copper content using bismuth-borate glasses based on 70 wt% Bi_2O_3–29.5 wt% B_2O_3–0.5 wt% Dy_2O_3 at T = 298 K (P = 0.95, n = 5)

Object of analysis	Analysed component	Calibration dependence equation in the form $\ln[(I_o/I) - 1] = a + b \cdot [\ln C(Cu)]$[a]	R^2	$C(Cu)_{min}$ (wt%)
CuO	Cu	$\ln[I_o/I - 1] = 5.01 + 1.03 \cdot \ln C(Cu)$	0.992	0.001

[a]I_o is the intensity of the luminescent radiation of Dy^{3+} in a glassy material having a composition of 70 wt% Bi_2O_3–29.5 wt% B_2O_3–0.5 wt% Dy_2O_3

The obtained results show that the dysprosium luminescence is almost completely quenched under the copper content of 0.06 wt% (Fig. 5b). Therefore, its luminescence determination is impossible at higher concentrations of quenchers. A procedure for the luminescence determination of copper content at room temperature using bismuth-borate glassy RMs doped with Dy_2O_3 was developed.

The analysis of the luminescence spectra of bismuth-borate glassy RMs at a temperature of 77 K shows that there exists a simultaneous quenching by Cu^{2+} ions of both Bi^{3+} and rare earth ionic luminescent radiation, making it difficult to use quantitative characteristics of the luminescence quenching process for determining the quencher element. Therefore, it seems reasonable to use room temperatures when carrying out luminescence analysis of a quencher element using bismuth-borate glassy RMs, at which the exciton transitions of Bi^{3+} ions insignificantly affect the luminescence process.

4 Conclusion

1. Bismuth-borate glassy RMs are found to luminesce at a temperature of 77 K in the same way as germanoeulytine crystals due to an exciton mechanism. The radiative process is characterised by the presence of a broad spectral luminescence band associated with the lowest permissible $^3P_1 \rightarrow {}^1S_0$ transition of Bi^{3+} ions. At room temperature, no luminescent radiation of the Bi^{3+} ions in the materials has been observed.
2. The photoluminescence process of bismuth-borate glassy RMs doped with the oxides of transition and rare-earth elements at 77 and 298 K has been studied. It is established that the presence of luminescence quenching elements (Cu, Co, Fe) in the composition of bismuth-containing glassy RMs leads to the quenching of the exciton band of intrinsic radiation of Bi^{3+} ions at 77 K. In the series of studied quencher elements Cu–Co–Fe, the greatest quenching effect of the intrinsic luminescence of bismuth ions (by a factor of around 1.5) has been observed for materials containing Cu^{2+} ions. The obtained dependencies of the

intensity of the luminescent radiation of the Bi^{3+} ions on the transition element ion content are described using the Stern-Volmer equation. In these equations, the coefficients for the studied transition elements are different, allowing the content of luminescence quenching elements to be determined for different transition element ions.

3. It is established that the luminescence of bismuth-borate glassy RMs activated by rare-earth ions at a temperature of 77 K is characterised by the presence of intrinsic exciton luminescence of glasses (Bi^{3+}) and the emission bands of the activator element ion caused by f-f transitions between the terms of rare earth ions.
4. It is established that, given the joint presence of REI (Dy) and transition element ions (Cu) in bismuth-borate glassy RMs at 77 K, the luminescence of lanthanide and intrinsic exciton radiation of glasses (Bi^{3+}) are quenched by transition element ions.
5. A measurement procedure has been developed for determining the low-temperature luminescence of copper in Bi-HTS of $Bi_2Sr_2CuO_6$ composition, which has a relative error of no more than 0.05 and a detection limit of 0.001 wt%. It is established that a determination of the copper content in Bi-HTS, which contain (along with copper) elements not involved in quenching Bi^{3+} ions, can be carried out using RMs doped solely with copper oxides. The correctness of the developed procedure has been verified using XRF and AA analyses.

References

1. Kutivitskij VA, Borisova VV, Mironova EV, Maruf M (2015) Development of the method of synthesis for polyfunctional vitreous samples on the basis of glassy flux in Bi_2O_3-B_2O_3 system. Ref Mater 4:34–45 (In Russ.)
2. GOST 8.531-2002 (2002) State system for ensuring the uniformity of measurements. Reference materials of composition of solid and disperse materials. Ways of homogeneity assessment. IPK Izdatelstvo standartov, Moscow, p 12 (in Russ.)
3. R 50.2.031-2003 (2003) State system for ensuring the uniformity of measurements. Reference materials for the composition and properties of substances and materials. Procedure for assessing stability characteristics. IPK Izdatelstvo standartov, Moscow, p 7 (in Russ.)
4. Pustovarov VA (2016) Rare-earth ion spectroscopy. Electronic publication. UrFU, Ekaterinburg, p 69. https://www.stady.urfu.ru
5. Shulgin BV, Polupanova TI, Kruzhalov AV et al (1992) Bismuth Orthogermanate, Vneshtorgizdat, Sverdlovsk, p 170 (in Russ.)
6. Marfunin AS (1985) Spectroscopy, luminescence and radiation centers in minerals. Nedra, Moscow, p 320 (in Russ.)
7. Poluektov NS (1990) Spectroscopy in coordination and analytical chemistry. Naukova Dumka, Kiev, p 119 (in Russ.)
8. RMG 61-2010 (2012) Accuracy, trueness and precision measures of the procedures for quantitative chemical analysis. In: Methods of evaluation. Standartinform, Moscow, p 58 (in Russ.)

Participation of the All-Russian Research Institute of Physical-Technical and Radio-Technical Measurements (VNIIFTRI) in International Comparisons in the Field of Physicochemical Measurements

Vladimir I. Dobrovolskiy, Narine G. Oganyan, Sergey V. Prokunin and Alexey A. Stakheev

Abstract The present article describes the participation of the All-Russian Research Institute of Physical-technical and Radio-technical Measurements (VNIIFTRI) in international comparisons in the field of physicochemical measurements, particularly in the field of pH and pX measurements, and in the field of the determination of inorganic component content in different matrices. New CMCs of VNIIFTRI entered into the BIPM database are also given in the paper.

Keywords Key comparison · Measurement standard · Standard-titre · Buffer solution · Reference material · pH · pX · Electrochemical analysis · Inorganic analysis · CMC

The Research Department of Physicochemical and Electrical Measurements (NIO-6) of the All-Russian Research Institute of Physical-technical and Radio-technical Measurements (VNIIFTRI), being one of the leading metrological centres in Russia, takes an active part in international and regional key, supplementary and pilot comparisons with the purpose of confirming its calibration and measurement capabilities (CMCs), developing a reference base and ensuring the uniformity of physicochemical measurements in Russia at the level of current international requirements. These comparisons were conducted in the framework of Working groups of the Consultative Committee on the Amount of Substances of the International Bureau of Weights and Measures (CCQM BIPM) on a par with leading international institutions.

Thus far, NIO-6 has successfully participated in international key, supplementary and pilot comparisons with the following State Primary Measurement Standards of VNIIFTRI:

V. I. Dobrovolskiy · N. G. Oganyan (✉) · S. V. Prokunin · A. A. Stakheev
All-Russian Research Institute of Physical-Technical and Radio-Technical Measurements (VNIIFTRI), Moscow Region, Russia
e-mail: mera@vniiftri.ru

S. Medvedevskikh et al. (eds.), *Reference Materials in Measurement and Technology*,
https://doi.org/10.1007/978-3-030-32534-3_19

- State Primary Measurement Standard of pH in aqueous solutions (GET 54-2011);
- State Primary Measurement Standard of ion activity (pX) in aqueous solutions (GET 171-2011);
- State Primary Measurement Standard for the units of mass fraction and mass (molar) concentration of inorganic components in aqueous solutions based on gravimetric and spectral methods (GET 217-2018).

Since 2017, the researchers carrying out works on improving the State Primary Standard for the units of dispersed parameters of aerosols, suspensions and powdery materials GET 163 (2015–2017 years) have actively joined international comparison studies and are now planning to participate in two projects in the framework of the Working Groups on gas and inorganic analysis (GAWG and IAWG) of CCQM BIPM:

- CCQM-K150-P189-2017 Comparison for particle number and charge concentration;
- CCQM-P194-2018 Pilot comparison on the determination of number concentration of colloidal nanoparticles.

In order to confirm the equivalence of measurements performed using two GETs, which are located at VNIIFTRI, NIO-6 has organised pilot comparisons within the framework of COOMET (Euro-Asian Cooperation of National Metrological Institutions), acting as coordinators of the following projects:

- 573/RU/12 Pilot comparisons in the field of measuring the number concentration of air ions (GET 177-2010 State primary standard of the units of the volume density of the electric charge of ionised air and the number concentration of air ions);
- 583/RU/12 Determination of the wide range of concentrations of dissolved oxygen in water (GET 212-2014 State primary standard of the unit of mass concentration of oxygen and hydrogen in liquid media).

The COOMET project 573/RU/12 has already been completed. A final report is being prepared to finalize work on the project 583/RU/12.

1 State Primary Standard of pH in Aqueous Solutions (GET 54-2011) and State Primary Standard of Ion Activity (pX) in Aqueous Solutions (GET 171-2011)

The State Primary Standards of pH and the State Primary Standard of pX in aqueous solutions have been located at VNIIFTRI since as early as 1990s.

Since July 2013, both measurement standards have been maintained by a new custodian, Sergey V. Prokunin, Cand. Sci. (Engineering). During 5 years of his work, VNIIFTRI took part in 6 international comparisons of its calibration and measurement capabilities with the application of GET 54-2011 and GET 171-2011. All these comparisons were conducted within the framework of the Working Groups on Electrochemical and Inorganic Analysis (EAWG and IAWG) of CCQM BIPM:

- CCQM-K48.2014. Key comparison. Assay of Potassium Chloride [1];
- CCQM-P152.2014. Pilot Study. Acidity function of phthalate buffer in water/ethanol mixture (mass fraction 50%), Draft A;
- CCQM-K99.2015. Key comparison on pH of an unknown phosphate buffer [2];
- APMP.QM-K91. 2016 APMP comparison on pH measurement of phthalate buffer [3];
- CCQM-K18.2016. Key comparison pH of carbonate buffer. Planned;
- CCQM-K19.20XX Key Comparison on Borate Buffer. Planned.

In addition, a series of comparisons was organised within the framework of COOMET (TC1.8/SC3 "Electrochemistry"): COOMET 655/RU/15 "Pilot study on pH measurement of phosphate buffer (pH ~ 7.0)". The COOMET 655/RU/15 comparison process was repeatedly reported at the meetings of COOMET TC 1.8 and CCQM EAWG. At present, this work is at the stage of completion.

Figures 1 and 2 show the results of VNIIFTRI participation in the international comparisons within the framework of CCQM BIPM—APMP.QM-K91.2016 (the latter of the completed comparisons of pH given above) [3]. As can be seen from the figures, NIO-6 conducted the measurements at a high level. The project was participated by the following organisations: ROChem-LIPI (Indonesia), VMI (Vietnam), CMI (Slovakia), VNIIFTRI (Russia), Tubitak Ume (Turkey), CENAM (Mexico), INACAL (Peru), NML-SIRIM (Malaysia), BelGIM (Belarus), GLHK (Hong Kong).

The positive results of the international key and supplementary comparisons not only have confirmed the equivalence of the results obtained using the State Primary Standard GET-54 to those achieved in the leading countries in this field (thus confirming the already available five CMCs in the BIPM database), but has also allowed another CMC on pH determination to be added with the corresponding measurement uncertainties, at a 95% confidence level and a coverage factor of k = 2, The above-mentioned CMCs of VNIIFTRI are provided below:

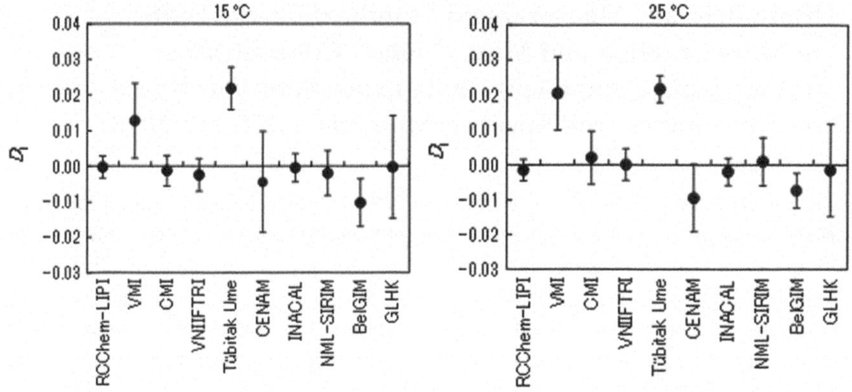

Fig. 1 Degree of equivalence Di and the expanded uncertainty Ui at temperatures of 15 and 25 °C

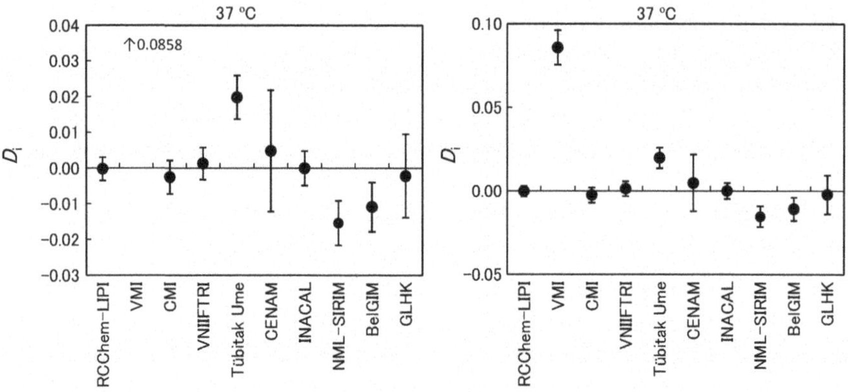

Fig. 2 Degree of equivalence Di and the expanded uncertainty Ui at a temperature of 37 °C

- pH from 3.533 to 3.561 with an expanded uncertainty of 0.004;
- pH from 4.001 to 4.009 with an expanded uncertainty of 0.004;
- pH from 6.861 to 6.869 with an expanded uncertainty of 0.004;
- pH from 6.86 to 7.41 with an expanded uncertainty of 0.02;
- pH from 9.176 to 9.184 with an expanded uncertainty of 0.008;
- pH from 3.80 to 4.20 with an expanded uncertainty of 0.02.

Today, on the basis of NIO-6, the production of Class 1 and Class 2 standard-titres and buffer solutions for pH and pX measurements has been organized, along with the production of ORP standard-titres, acidity measures and working standards of ion activity in aqueous solutions. Due to the confirmation of its measurement capabilities at the world level, VNIIFTRI can confidently declare the high quality of its products.

2 State Primary Measurement Standard for the Units of Mass Fraction and Mass (Molar) Concentration of Inorganic Components in Aqueous Solutions Based on Gravimetric and Spectral Methods (GET 217-2018)

In the period from 2013 to 2017, a group of NIO-6 researchers developed the State primary standard for the units of mass fraction and mass (molar) concentration of inorganic components in aqueous solutions based on gravimetric and spectral methods GET 217-2018. Alexey A. Stakheev, Cand. Sci. (Engineering), was appointed as a custodian. In the process of developing the standard, the researchers expressed their readiness to take part in several international comparisons within the framework of the CCQM BIPM:

- SIM.QM-S7 Supplementary Comparison. Trace Metals in Drinking Water [4];
- APMP.QM-S10 Supplementary Comparison. Elements in Food Supplement. Measurements completed;
- EURAMET.QM-S11 Supplementary Comparison. Elements in River Water. In progress;
- CCQM-K128. Key comparison. Measurement of Heavy Metals and Organo-Tin in Leather Powder. In progress;
- CCQM-K145 Key comparison. Toxic and essential elements in bovine liver powder. Planned;
- CCQM-K155 and P196: Elements in seawater. Planned.

Out of the six indicated comparison projects, SIM.QM-S7 Supplementary Comparison "Trace Metals in Drinking Water", which were announced in October 2015 and continued until the end of 2017, were completed. The aim of the project was to assess the analytical capabilities for an accurate determination of trace metal quantity in drinking water at National Metrology Institutes (NMIs) and Designated Institutions (DIs) of the BIPM member countries. The following 16 NMIs/DIs from 15 countries participated in the project: Water (NRC (Canada), CNM (Mexico), INTI (Argentina), LTU (Uruguay), VNIIFTRI (Russia), BIM (Bulgaria), FEI (Finland), LCM (Costa Rica), The Bureau of Standards Jamaica (Jamaica), NIM (Thailand), NIM (Romania), NMI (Colombia), UNIIM (Russia), GSCL (Greece), IBM (Bolivia), Trinidad and Tobago Bureau of Standards (Trinidad) [4]. The coordinators allowed NMIs/DIs to choose measurement methods. According to the reports received, most NMIs/DIs used inductively coupled plasma mass spectrometry (ICP-MS) in their studies. VNIIFTRI (at that moment, GET 217-2018 was in the status of the developed standard) announced its participation to determine all of four offered metals.

Upon the completion of the work, the project SIM.QM-S7 was approved by the members of CCQM IAWG. Figures 3 and 4 provide the results on the participation of VNIIFTRI in the SIM.QM-S7 Supplementary Comparison "Trace Metals in Drinking Water" [4].

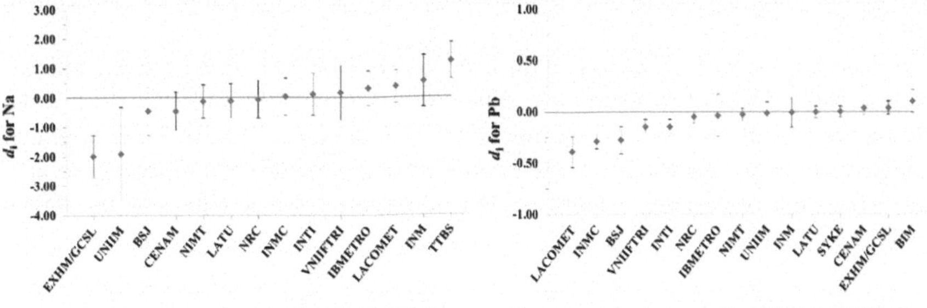

Degree of equivalence for SIM.QM-S7 Na based on DSL calculation

Degree of equivalence for SIM.QM-S7 Pb based on DSL calculation

Fig. 3 Degree of equivalence between sodium and lead measurements performed using GET 217 and measurement results of various countries in the SIM.QM-S7 comparison

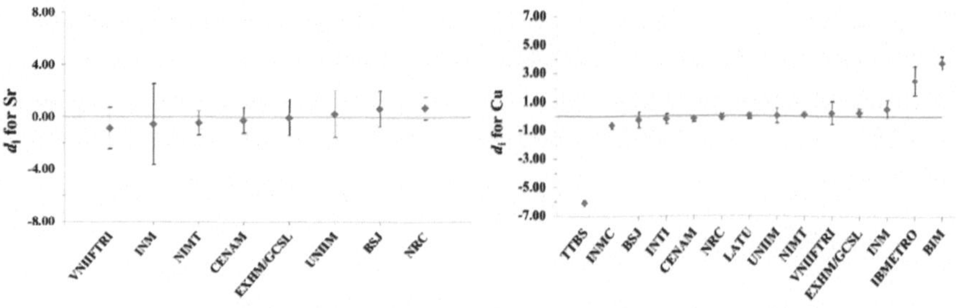

Degree of equivalence for SIM.QM-S7 Sr based on DSL calculation

Degree of equivalence for SIM.QM-S7 Cu based on DSL calculation

Fig. 4 Degree of equivalence between strontium and copper measurements performed using GET 217 and measurement results of various countries in the SIM.QM-S7 comparison

Taking into account the positive experience of its participation in the project (at the beginning of 2018), VNIIFTRI claimed 4 new CMCs in the BIPM database. The CMCs for determining the mass fraction of the trace metals in drinking water with the corresponding measurement uncertainties, at a 95% confidence level and a coverage factor of k = 2, are listed below:

Cu (Copper) 1–100 μg/kg, with an expanded uncertainty of 0.11–8 μg/kg;
Sr (Strontium) 1–100 μg/kg, with an expanded uncertainty of 0.05–3.5 μg/kg;
Pb (Lead) 0.1–100 μg/kg, with an expanded uncertainty of 0.15–10 μg/kg;
Na (Sodium) 0.5–50 mg/kg, with an expanded uncertainty of 0.035–2.5 mg/kg.

The recognition of the measurement capabilities of VNIIFTRI (measurement standard GET 217-2018) in the indicated ranges with the specified uncertainties will allow the laboratory to launch the production of relatively inexpensive reference materials of copper, strontium, lead and sodium in aqueous solutions traceable to the CMCs of KCDB BIPM database. Positive results in other open comparison projects will permit VNIIFTRI to increase the number of CMCs with the further prospect of expanding the list of developed reference materials of inorganic components in liquids.

In conclusion, VNIIFTRI has confirmed and expanded the measurement and calibration capabilities of measurement standards GET 54-2011 and 171-2011 at the international level. In addition, the functionality of the GET 217-2018 measurement standard has been confirmed by the results of one completed comparison project out of 6 currently running, which fact is evidenced by 4 new CMCs in the BIPM database.

References

1. Liandi MA et al (2016) CCQM-K48.2014: assay of potassium chloride. Metrologia 53:08012. https://doi.org/10.1088/0026-1394/53/1a/08012
2. CCQM-K99 Final Report, Chemistry and Biology, list of comparison, KCDB, BIPM. https://www.bipm.org/utils/common/pdf/final_reports/QM/K99/CCQM-K99_Final_Report.pdf
3. Hioki A et al (2017) Final report of the key comparison APMP.QM-K91: APMP comparison on pH measurement of phthalate buffer. Metrologia 54:08002. https://doi.org/10.1088/0026-1394/54/1a/08002
4. Yang L et al (2018) Final report of the SIM.QM-S7 supplementary comparison, trace metals in drinking water. Metrologia 55:08002. https://doi.org/10.1088/0026-1394/55/1a/08002

Quality Improvement of Commercially Available Certified Reference Gas Mixtures

Leonid A. Konopelko, Anna V. Kolobova and Olga V. Fatina

Abstract In Russia, the metrological traceability of certified reference materials for the composition of gas mixtures in cylinders under pressure produced by the manufacturers of certified reference materials is carried out in accordance with GOST 8.578-2014. Certified reference materials for the composition of gas mixtures in cylinders under pressure are applied in type approval testing of measuring instruments, verification, calibration and graduation of gas analytical measuring instruments intended for controlling fire and explosion hazardous gases and vapours, harmful components in the atmospheric and workplace air, emissions of vehicles and industrial enterprises, control of technological processes and quality of hydrocarbon products, etc. In this connection, ensuring the quality of commercially available certified reference materials, which number has recently amounted to about 100,000 cylinders with gas mixtures per year, is an important task. In this paper, we suggest a set of measures aimed at ensuring the quality of certified reference gas mixtures in cylinders under pressure, mass-produced by the manufacturers of certified reference materials: mandatory accreditation of reference material producers (RMPs) for compliance with the requirements of GOST ISO Guide 34-2014 and GOST ISO Guide 35-2015; constant participation of RMPs in proficiency testing programmes through interlaboratory comparisons; updating and revision of the existing set of standards defining the requirements for the entire life cycle of certified reference gas mixtures in cylinders under pressure; improvement of the method for reference material certification by calculating the value of their expanded uncertainty and introducing of a new, "technological reserve", factor.

Keywords Metrology · Certified reference material · Gas mixture · State primary measurement standard · Secondary measurement standard · Working measurement standard · Traceability · Quality control

---

L. A. Konopelko · A. V. Kolobova · O. V. Fatina (✉)
D.I. Mendeleyev Institute for Metrology (VNIIM), Saint Petersburg, Russia
e-mail: fhi@b10.vniim.ru

S. Medvedevskikh et al. (eds.), *Reference Materials in Measurement and Technology*,
https://doi.org/10.1007/978-3-030-32534-3_20

1 Introduction

In accordance with the Russian state hierarchy scheme instruments for measuring the content of components in gaseous mediums [1], certified reference materials for the composition of gas mixtures in cylinders under pressure (hereinafter referred to as CRMs) are working measurement standards, and, in fact, according to [2, 3], CRMs are material measures for measuring instruments, which are used for tests to approve the type of measuring instruments, verification, calibration and graduation of gas-analytical measuring instruments. The extensive use of certified reference gas mixtures in cylinders under pressure during the production and control of measuring instruments is evidenced by the annual release of about 100,000 gas mixture cylinders only in Russia.

In Russia, the metrological traceability of CRMs for the composition of gas mixtures in cylinders under pressure is implemented according to the scheme provided in [1] (see Fig. 1). The reproduction, storage and transfer of the unit of mole fraction of components in gaseous media is carried out with the help of the State primary measurement standard of the units of mole fraction, mass fraction and mass concentration of components in gaseous and gas condensate media GET 154-2016[1] (hereinafter referred to as GPE GET 154) functioning at the D.I. Mendeleev Institute

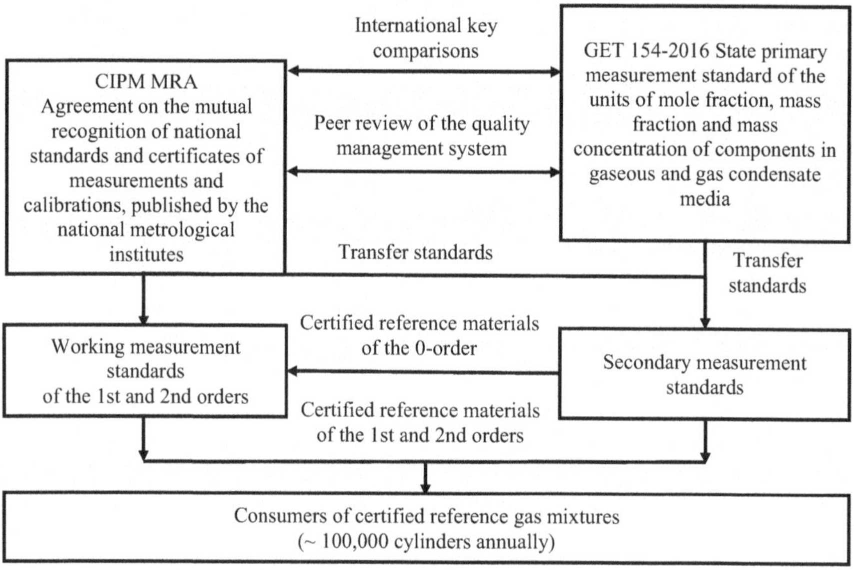

Fig. 1 Metrological traceability of certified reference gas mixtures in cylinders under pressure in accordance with [1]

[1]GET 154-2016 State primary measurement standard of the units of mole fraction, mass fraction and mass concentration of components in gaseous and gas condensate media. VNIIM. Available at: http://fhi.vniim.ru/struct/standards/get-154.

for Metrology (VNIIM). Produced using the GPE GET 154 standard equipment, certified reference gas mixtures in cylinders under pressure perform the function of transfer standards according to [1], exhibit the highest accuracy in the country and are used to transfer the unit of mole fraction of components to secondary and working measurement standards, which operate at enterprises producing certified reference gas mixtures in cylinders under pressure for consumers.

2 Literature Review

In accordance with the Russian regulatory legal acts [4–7], measuring instruments used for ensuring the uniformity of measurements, including measures of any physical quantity [2, 3], should undergo:

- tests for the purpose of approving the type of measuring instruments and its registration in the Federal Information Fund for Ensuring the Uniformity of Measurements;
- primary and periodic verification or calibration during operation by accredited legal entities or individual entrepreneurs according to [8].

These requirements are partially applied to CRMs performing the role of material measures for measuring instruments. Currently, all such CRMs are supposed to pass only type approval testing and type registration in the Federal Information Fund for ensuring the uniformity of measurements [4–7]. No obligatory procedures, similar to those applied to material measures, are used for these CRMs after their production and during operation.

Certified reference gas mixtures are applied in type approval testing of measuring instruments; verification, calibration and graduation of gas analytical measuring instruments intended for controlling fire and explosion hazardous gases and vapours, harmful components in the atmospheric and workplace air, emissions of vehicles and industrial enterprises; control of technological processes and quality of hydrocarbon products, etc. In this connection, ensuring the quality of commercially available certified reference gas mixtures is an important task.

3 Materials and Methods

Experience in the production of CRMs shows that their metrological traceability [1] cannot be considered as a sufficient condition for ensuring the quality of commercially available CRMs. This is explained by the complexity and multi-stage character of the process [9] of producing certified reference gas mixtures in cylinders under pressure. The procedure for transferring the unit of mole fraction of components in gas mixtures from the GPE GET 154 is carried out by annual calibration or verification of gas mixing and analytical equipment constituting the secondary and working measurement standards [1] used in reference material certification. At the same time, this procedure does not cover all the stages of the complex process of CRM production implemented at respective enterprises; therefore, it cannot guarantee the quality [9] of all commercially produced CRM units at any time period within the established shelf life.

The criterion for compliance with the required level of quality of CRMs is the compliance of produced CRMs with the established characteristics in the type description of a CRM [4–7].

In the general case, the value of the combined standard uncertainty [10] of a CRM (u_{CRM}) over its entire lifetime (t_{add}) is a function of many variables:

$$u_{CRM} = f(u_{cert}, u_{tadd}), \text{ where} \tag{1}$$

$$u_{\text{cert}} = f(u_{SS,WS,TS}, u_{\text{meas}}), \tag{2}$$

$$u_{tadd} = f(u_{\text{phs}}, u_{\text{chs}}, u_{\text{chmut}}, u_{\text{exp}}, u_{\text{imp}}), \tag{3}$$

where

u_{cert} is the standard uncertainty of the certified value of the mole fraction of the analysed component in the CRM;
$u_{SS,WS,STS}$ is the standard uncertainty associated with the transfer of the mole fraction unit of components in gas mixtures from the GPE GET 154 via the transfer standards to the secondary measurement standard or working measurement standard used to certify this reference material;
u_{meas} is the standard uncertainty of measurement results obtained during the reference material certification;
u_{tadd} is the standard uncertainty due to various physical and chemical processes occurring during the entire period of the CRM use;
u_{phs} is the standard uncertainty due to the process of physical sorption of gases during filling the cylinder;
u_{chs} is the standard uncertainty caused by the process of chemical sorption of gases during storage of the gas mixture in the cylinder;
u_{chmut} is the standard uncertainty due to the mutual influence of gas components, including the interaction of impurity components with the analysed component in the gas mixture;
u_{exp} is the standard uncertainty due to a change in the state of the gas mixture caused by changes in the pressure of the gas mixture in the cylinder during the expenditure of the CRM;
u_{imp} is the standard uncertainty due to the presence of impurities that act as interfering components for the verified (calibrated) gas analyser.

For the main condition of the CRM applicability to be satisfied, the following condition must be ensured

$$U_{\text{ti(where ti} \leq \text{tadd)}} \leq U_{\text{est}}, \quad (4)$$

where

$U_{\text{ti(where ti} \leq \text{tadd)}}$ is the value of the expanded uncertainty ($k = 2$) of the analysed component in the CRM at any time during its use within the established shelf life;
U_{est} is the value of the expanded uncertainty ($k = 2$) of the analysed component in the CRM, established through type approval testing and specified in the CRM type description.

In addition, different CRM units representing gas mixtures of the same kind, composition and nominal value of the analysed components, produced by different RMPs, must be comparable:

$$|X_{\text{CRM1}} - X_{\text{CRM2}}| \leq k \cdot \sqrt{u^2_{\text{A(CRM1)}} + u^2_{\text{A(CRM2)}}}, \quad (5)$$

where

X_{CRMi} is the measured value of the mole fraction of the analysed component in the i-th CRM, submitted for inspection by the i-th RMP, where $i = 1$ or 2,
$u_{A(CRMi)}$ is the type A standard uncertainty obtained during certification of the i-th reference material using the secondary or working measurement standard of the i-th RMP, where $i = 1$ or 2,
k is the coverage factor, $k = 2$.

The comparability of CRMs is monitored using the high-precision gas analytical equipment of the GPE GET 154 (the uncertainty value is negligible).

The ISO standards TC 158 “Analysis of gases” and TC 193 “Natural gas” suggest the following methods for estimating the uncertainty u_{tadd} (see formula (3)):

- verification of gravimetric gas mixtures;
- assessment of the quality of the inner surface of the cylinder;
- reference material certification followed by the analysis of the results in two or three stages at considerable time intervals;
- control of the gas mixture composition during the reduction of pressure in the cylinder, etc.

It should be noted that the application of all these methods for estimating the uncertainty u_{tadd} requires meticulous and time-consuming work. Thus far, CRMs have been mass-produced based on the principle of minimising the labour intensity of the manufacturing process. At the same time, it has been assumed that the extensive experience accumulated in the field of CRM production would provide the required level of the quality of gas mixtures in terms of their conformity with the expected characteristics. Therefore, the process of reference material certification has been so far based on the use of both standard gas mixing and analytical equipment included into the secondary and working measurement standards, as well as typical certification algorithms. The certified value of the mole fraction of the analysed component in a gas mixture has been assigned the expanded uncertainty value ($k = 2$), which was established during type approval testing and specified in the type description of respective CRMs.

4 Results and Discussion

The analysis of approaches aimed at improving CRM quality, which can be introduced across the entire industry of reference material production, has identified the following set of actions:

1. Mandatory accreditation of all RMPs for their compliance with the requirements of GOST ISO Guide 34-2014 [11] and GOST ISO Guide 35-2015 [12]. This step will require the implementation of the quality management system at enterprises producing reference materials.
2. Constant participation of RMPs in proficiency testing programmes through interlaboratory comparisons, including the participation of CIS and European enterprises [13].
3. Updating and revision of the existing set of standards defining the requirements for the entire CRM life cycle (i.e. CRM development, type approval tests, production, logistics, production control and use), taking into account the mandatory estimation of uncertainty during reference material certification, as well as the implementation of a new set of standards and the requirements of the ISO TC 158 and TC 193;
4. Introduction of the developed requirements into practice, taking into account the updated set of standards:
   - use of cylinders, including specialised ones, for various kinds of certified reference gas mixtures;
   - use of pure gases;
   - use of gas mixing and analytical equipment included into the secondary and working measurement standards of RMPs.
5. Centralised production of cylinders, including specialised ones, as well as pure gases and substances.
6. Introduction of the mandatory requirement to use only labelled cylinders for the same kinds of gas mixtures, preferably on a returnable basis.
7. Implementation of a new method for certifying reference materials: substitution of the existing method of assigning the value of expanded uncertainty of the certified value of the analysed component established during CRM type approval testing [4–7] with the method of calculating the expanded uncertainty of the reference material certification using the technological reserve factor described below.

Table 1 Uncertainty budget for the certification of reference gas mixtures using the gas analytical equipment of the working measurement standard in accordance with GOST ISO Guide 35-2015 [12]

No	Sources of uncertainty	Stage of determination
1	u_1 uncertainty of characterisation, including the following components	In the course of reference material certification
1.1	Uncertainty of the certified value of the transfer standard	Given in the passport of the transfer standard
1.2	Uncertainty associated with comparison	In the course of reference material certification
2	u_2 uncertainty associated with short-term stability	In the course of reference material certification
3	u_3 uncertainty associated with long-term stability	In the course of testing the reference material for the purposes of type approval
4	u_4 uncertainty associated with heterogeneity	In the course of testing the reference material for the purposes of type approval and certification

We suggest a new method of reference material certification, which conforms to internationally recognised approaches and is based on the introduction of the technological reserve factor.

It should be noted that international approaches to the reference material production are fundamentally different from that used in Russia. In accordance with the GOST ISO Guide 34-2014 [11] and GOST ISO Guide 35-2015 [12], when reference materials are produced in single units, each reference material undergoes the certification process, including the assessment of its standard deviation, stability and homogeneity. Based on the measurement results, the uncertainty budget is compiled, which generally comprises uncertainty of characterisation, as well as uncertainties due to CRM instability and heterogeneity. Table 1 presents an example of the uncertainty budget complied during gas mixture certification using the gas analytical equipment of the working measurement standard. On the basis of measurement results, the certified value of the mole fraction of the analysed components in the gas mixture under study is determined for each CRM, along with the expanded uncertainty calculated under the coverage factor of $k = 2$.

It should be noted that the characterisation uncertainty u_1, indicated in Table 1 corresponds to the certification uncertainty u_{cert} presented in formula (2). At the same time, the uncertainties u_2, u_3 and u_4 caused by instability and heterogeneity (Table 1) only partially take into account the sources of uncertainty used for the estimation of uncertainty u_{tadd} in formula (3).

In order to reliably account for all the sources of uncertainty u_{tadd}, as well as those caused by the human factor, we suggest the introduction of a so-called 'technological reserve' factor. This factor is supposed to allow for a number of unaccounted influences that may arise during the multi-stage technological process of reference material production:

$$K = f(A_1, A_2, A_3, A_4, \ldots B_1, B_2, B_3, B_4), \quad (6)$$

where

A_i are unaccounted influences that may arise at each stage of the multi-stage technological process of reference material production (preparation and selection of cylinders and pure gases, filling cylinders, etc.),
B_i is the human factor (competence of employees, their adherence to all documented procedures, etc.).

Therefore, we suggest that reference material certification should be carried out in accordance with the following considerations:

- the set of statistical data should be sufficient to allow for all uncertainty sources;
- the certification of each reference material unit should involve the calculation of the expanded uncertainty value under the coverage factor of $k = 2$;
- the result of certification is considered positive provided that the calculated value of the expanded uncertainty taking into account the technological reserve factor does not exceed the value of the expanded uncertainty with the coverage factor $k = 2$, which was established during type approval testing and specified in the CRM type description [4–7].

The condition for checking the result of reference material certification using the technological reserve factor takes the following form:

$$U_{calc} \cdot K \leq U_{est}, \quad (7)$$

$$U_{calc} = f(u_1, u_2, u_3, u_4), \quad (8)$$

where

U_{calc} is the calculated value of the expanded uncertainty ($k = 2$) obtained when certifying a particular reference material unit;
U_{est} is the value of the expanded uncertainty ($k = 2$) established during type approval testing and specified in the CRM type description;
K is the technological reserve factor established for the representative groups of gas mixtures.

Considering that the entire range of CRMs, produced in the amount of 100,000 cylinders, comprises over 10,000 gas mixture kinds, all of which differ in the component composition, quantitative characteristics of each component in the mixture, deviations for the preparation of each component and the values of the expanded uncertainty of each component in the mixture, the use of a single factor characterising the technological reserve for all kinds of mixtures appears to be impossible.

Therefore, it seems reasonable to select a few groups that would be representative of the entire set of gas mixture kinds. Such a selection should be carried out on the basis of the following principles:

- use of the same type of cylinders, valves and other gas fittings;
- use of the same type of methods for the preparation of cylinders for gas mixtures based on reactive gases, inert gases, permanent gases, micro-concentrations of the analysed components, etc.;
- use of versatile preparation instruments;
- use of the same type of gas mixing and analytical equipment comprising secondary and working measurement standards.

Each representative group of gas mixtures would require its own technological reserve factor, which would be taken into account in measurement methods used by various producers of certified reference gas mixtures, as well as in the quality control of CRMs using GPE GET 154.

An approach of merging a number of gases into representative groups is also being discussed by the Consultative Committee for the Amount of Substance of the International Bureau of Weights and Measures within the Working Group on Gas Analysis. Thus, a number of gases and respective gas mixtures have been allocated into a so-called "flexible" group (flexible gases). The results of international interlaboratory comparisons for one gas from this "flexible" group can be extended to other gases from this group, thus confirming the existing calibration and measurement capabilities for all these gases.

In conclusion, the competence of RMPs, as well as the compliance of CRMs for the composition of gas mixtures utilised in Russia with strict contemporary requirements, can only be enhanced using an integrated approach aimed at ensuring the quality of commercially available CRMs. This approach includes the improvement of procedures for certifying reference materials, as well as those used by authorised organisations to monitor the quality of reference material production process at enterprises manufacturing CRMs. In addition, quality management systems meeting the requirements of GOST ISO 34 2014 [11] should be introduced by all RMPs.

Author Contribution Statement All authors have made an equivalent contribution to the preparation of the article for publication.

References

1. GOST 8.578-2014 (2014) State system for ensuring the traceability of measurements. State hierarchy scheme for measuring instruments of the content of components in gaseous mediums. Standartinform, Moscow (In Russ.)
2. RMG 29-2013 (2014) State system for ensuring the uniformity of measurements. Metrology. Basic terms and definitions. Standartinform, Moscow (In Russ.)
3. VNIIM, BelGIM (2010) International vocabulary of metrology: basic and general concepts and associated terms. NGO "Professional", St. Petersburg (In Russ.)
4. Federal law, On ensuring the uniformity of measurements, No FZ-102 of 26.06.2008. http://fundmetrology.ru/depository/01_npa/102-fz_2015.pdf (In Russ.)

5. Decree of the Ministry of Industry and Trade of the Russian Federation No 970 of 25.06.2013, On approval of the administrative regulations on the federal agency for technical regulation and metrology providing state services for the type approval of reference materials or measuring instruments. Moscow. http://www.fundmetrology.ru/depository/01_npa/pm1081_30112009.pdf (In Russ.)
6. Decree of the Ministry of Industry and Trade of the Russian Federation No 1081 of 30.11.2009, On approval of the procedure for testing reference materials or measuring instruments for the purposes of type approval; the procedure for type approval of reference materials or measuring instruments; the procedure for issuing certificates of reference material or measuring instrument type approval, establishing and changing the validity periods of the above-mentioned certificates and the interval between verifications of measuring instruments, requirements for type approval marks of reference materials or measuring instruments and marking procedure. Moscow. https://rg.ru/2018/11/09/minpromtorg-prikaz3249-site-dok.html (In Russ.)
7. Decree of the Ministry of Industry and Trade of the Russian Federation No 164 of 03.02.2015, On approval of the form of type-approval certificates of reference materials or measuring instruments. Moscow. http://base.garant.ru/70981308 (In Russ.)
8. Federal law, On accreditation in national accreditation system, No. 412-FZ of 28.12.2013. Moscow. http://rg.ru/2013/12/31/akkreditacia-dok.html (In Russ.)
9. Kolobova AV (2008) Research and development of methods for metrological control of industrially produced reference materials for the composition of gas mixtures. Dissertation. D.I. Mendeleyev Institute for Metrology (VNIIM), St. Petersburg (In Russ.)
10. GOST ISO Guide 34100.3-2017 (2017) Uncertainty of measurement. Part 3. Guide to the expression of uncertainty in measurement. Standartinform, Moscow (In Russ.)
11. GOST ISO Guide 34-2014 (2015) General requirements for the competence of reference material producers. Standartinform, Moscow (In Russ.)
12. GOST ISO Guide 35-2015 (2016) Reference materials. In: General and statistical principles for certification. Standartinform, Moscow (In Russ.)
13. GOST ISO/IEC 17043-2013 (2014) Conformity assessment. In: General requirements for proficiency testing. Standartinform, Moscow (In Russ.)

Application of Raman Spectroscopy in the Quality Control of Injectable Medicines

Maria P. Krasheninina, Maria Yu. Medvedevskikh, Ekaterina V. Galeeva and Roman R. Galeev

Abstract This paper presents the results of the identification procedure certification and subsequent quantitative determination of the active ingredients of two-component injectable medicines (active ingredient and solvent) using Raman spectroscopy. The main objective of the research was to select approaches for estimating the metrological characteristics of the measurement procedure, which include a consideration of the methodological parameters, along with a provision of the metrological traceability of measurement results to SI units. To this end, the GVET 176-1-2010 State Secondary Measurement Standard for units of mass fraction, mass (molar) concentration of components in solid and liquid substances and materials based on volumetric titration was used. The following substances were chosen as research objects for estimating the metrological characteristics of the measurement procedure: ascorbic acid, novocaine and sodium thiosulphate. The authors of the work have demonstrated the measurement-procedure certification results, whose accurate determination was confirmed by the results of interlaboratory comparisons. The obtained results confirmed the accuracy of the identification procedure and subsequent quantitative determination, hence its applicability for the determination of the active ingredients in two-component injectable medicines. In addition, the possibility of developing reference materials based on the medicines under study is indicated. Further development of this study may be directed at the development of an identification procedure and its certification, with subsequent quantitative determination of the active ingredients of injectable medicines having three components as well as those having a more complex composition.

Maria P. Krasheninina—ResearcherID: B 8302–2019
Maria Yu. Medvedevskikh—ResearcherID: G 6171–2017

M. P. Krasheninina (✉) · M. Yu. Medvedevskikh
Ural Scientific Research Institute for Metrology (UNIIM), Ekaterinburg, Russia
e-mail: krasheninina_m@uniim.ru

E. V. Galeeva (✉) · R. R. Galeev
Information and Methodological Center for Expert Evaluation, Recording and Analysis of Circulation of Medical Products of Roszdravnadzor, Moscow, Russia
e-mail: miniram532@gmail.com

S. Medvedevskikh et al. (eds.), *Reference Materials in Measurement and Technology*,
https://doi.org/10.1007/978-3-030-32534-3_21

Keywords Raman spectroscopy · Identification and authentication of medicines · Measurement procedure certification

1 Introduction

The quality of medicines used for treating people is vital for preserving their life and health. One of the main conditions for ensuring the quality of medicines is their identification and/or authentication, as well as the control of their active ingredients.

The study of current trends in the development of the latest analysis methods in the medical field shows that nondestructive methods that minimize analysis time are coming to the forefront in the vast majority of cases. In this connection, the Raman Spectroscopy (RS) method for routine measurements has the aforementioned advantages over traditional methods of wet chemistry, especially for injectable medicines, since it is non-destructive, contactless, rapid and does not require sample preparation [1]. For example, the application of the RS method for the characterization of aqueous ethanol solutions was described in [2] and in [3] for the simultaneous quantitative determination of glycerol and potassium acetate in an aqueous solution. The RS method is used for the qualitative and quantitative determination of the active ingredients of pharmaceutical products [4–12].

A Raman spectrum occurs when a substance is irradiated by a monochromatic laser beam of ultraviolet or visible range (the wavelengths ranging from ultraviolet to near infrared). When exposed to radiation, the molecules of the substance become polarized and scatter light in the range of wave numbers (from 2 to 4000 cm^{-1}). If the interaction of incident radiation quantum with a molecule in its ground or excited vibrational state is elastic, then the molecule energy state does not change and the scattered radiation frequency will be the same as the incident frequency (Raman/Rayleigh line). In the case of inelastic interaction, energy is exchanged between the emission quantum and the molecule, causing scattered radiation, which can be of greater or lesser frequency (anti-Stokes and Stokes lines, respectively), forming a Raman spectrum. The extreme sensitivity of Raman spectra to the nature of chemical bonds in organic molecules and polymeric materials, as well as in crystal lattices and clusters, determines the individuality of the spectrum of a particular substance [13]. A schematic representation of the transitions involved in the Raman scattering is shown in Fig. 1.

The present work was aimed at the development of an identification procedure followed by its certification, with a subsequent quantitative determination of the active ingredients of injectable medicines using the RS method, along with a confirmation of the metrological characteristics of the procedure by means of interlaboratory comparisons (hereinafter referred to as ICs).

The work was carried out by the Ural Scientific Research Institute for Metrology (UNIIM, Ekaterinburg) in conjunction with the Information and Methodological Center for Expert Evaluation, Recording and Analysis of Circulation of Medical

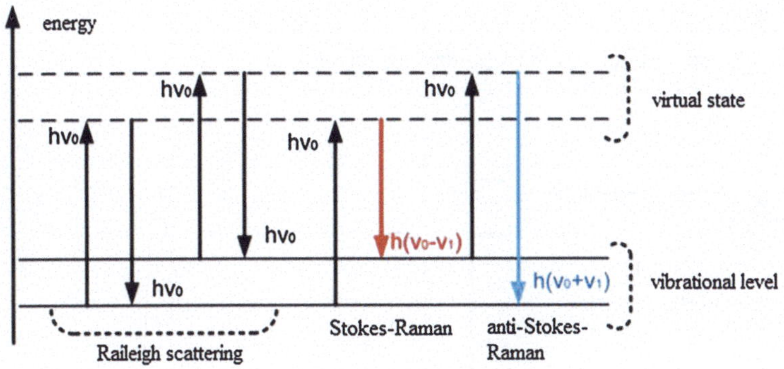

Fig. 1 Scheme showing transitions involved in Raman scattering [1]

Products of Roszdravnadzor, Department for the Development of Non-destructive Methods for Testing the Quality of Medicines (Kazan).

2 Certification of Measurement Procedure

The identification procedure, with subsequent quantitative determination of the active ingredients of injectable medicines using the RS method, is applicable to two-component injectable medicines consisting of a solvent and an active ingredient. The development of the measurement procedure is presented in [14].

Raman spectra were obtained using a Mini-Ram 532 Raman Spectrometer System (Fig. 2), which includes: an Enspectr R532++ (portable Raman-luminescence spectrometer having a spectral range from 200 to 4000 cm^{-1}) and a laser source having a wavelength of 532 nm. The spectra were analysed using the UniQue-Lab software and spectral database included in the Raman Spectrometer System (Fig. 3). In order to conduct the quality screening of injectable medicines within the state control

Fig. 2 General view of the Mini-Ram 532 Raman spectrometer system

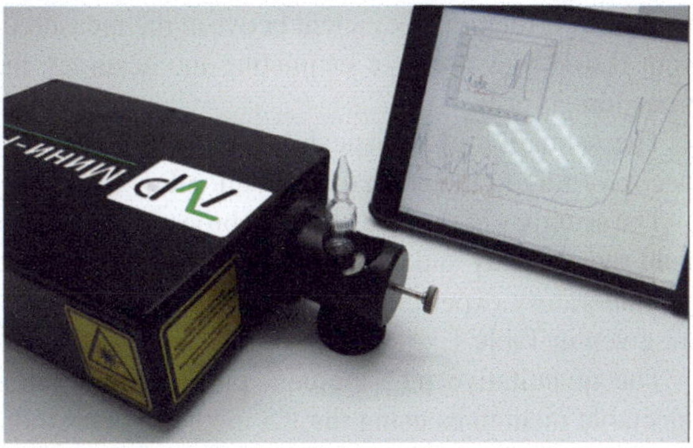

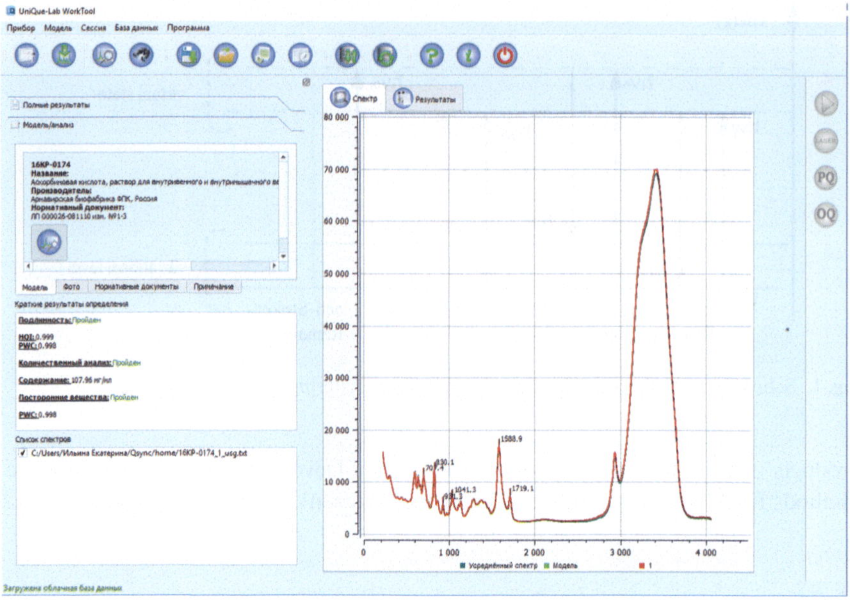

Fig. 3 The UniQue-Lab software interface

framework, a software program was developed by the Information and Methodological Center for Expert Evaluation, Recording and Analysis of Circulation of Medical Products of Roszdravnadzor. This program includes unique mathematical algorithms for signal processing, identification and quantification of injectable dosage forms [14].

Identification is a qualitative measurement procedure. The identification of medicines is carried out by comparing the Raman spectra of the test sample with similar spectra previously obtained from original pharmaceutical products and stored in the database.

There are currently no requirements for the development, design and legalisation of qualitative measurement procedures; therefore, a quantitative characteristic in the form of a correlation coefficient between the measured and stored spectra in a range from 0 to 1 was used for estimating the accuracy indicators of the measurement procedure.

The validation of statistical data when measuring the correlation coefficient for its conformity to the normal distribution law is carried out in accordance with [15].

The accuracy of the measurement procedure was estimated in the form of measurement repeatability and reproducibility in accordance with [16] by means of a quasi-interlaboratory experiment involving five laboratories. Metrological characteristics are given in Table 1.

The quantitative measurement procedure for the active ingredient content in injectable medicines using the RS method is based on a directly proportional relationship between the intensity (I) of spectral lines and the number of molecules (N)

Table 1 Measurement range, repeatability and reproducibility, in fractions

Measurement range of correlation coefficients	Measurement repeatability (the standard deviation of a single measurement under a set of repeatability conditions), σ_r	Measurement reproducibility (standard deviation under of a single measurement a set of reproducibility conditions), σ_R
From 0 to 1.0	0.008	0.013

per volume unit:

$$I = i \cdot k \cdot N, \tag{1}$$

where i is the intensity of scattered light per molecule;

k is the coefficient dependent on the conditions of the experiment (constant value for the measuring instrument used).

In the course of certifying the measurement procedure, the following medicines were used as evaluation samples: ascorbic acid, novocaine and sodium thiosulphate. The above-mentioned medicines were analysed using the GVET 176-1-2010 State Secondary Measurement Standard for units of mass fraction and mass (molar) concentration of components in solid and liquid substances and materials on the basis of volumetric titration. The use of GVET 176-1-2010, whose calibration and measurement capabilities were confirmed by means of participation in the CCQM-K130 [17] and CCQM-K149 [18] international key comparisons, meets the present-day requirement for the traceability of measurement results to SI units [19]. Examples of Raman spectra for ascorbic acid, novocaine and sodium thiosulphate are shown in Fig. 4.

The estimation of measurement precision in the form of repeatability and reproducibility was carried out in accordance with [16]. The estimation of measurement trueness was performed by evaluating the bias between the measurement results obtained using the measurement procedure and those obtained using GVET 176-1-2010.

In addition, a multi-factor experiment was conducted to assess the influence of methodological factors on measurement results [20].

The studied factors comprised: active ingredient content, packaging option (the volume of an ampoule), manufacturer, ambient temperature. The data was processed by applying regression analysis using the Microsoft Excel software package.

The limits of factor variation and the design of a multi-factor experiment are given in Tables 2 and 3, respectively.

In the course of processing the data of the multi-factor experiment, the following bias values were obtained:

– for 5 mg/cm^3 novocaine content

$$\Delta\Pi = 0.253(C - C_0) + 0.426(V - V_0) + 0.0001(P - P_0) + 0.108(T - T_0); \tag{2}$$

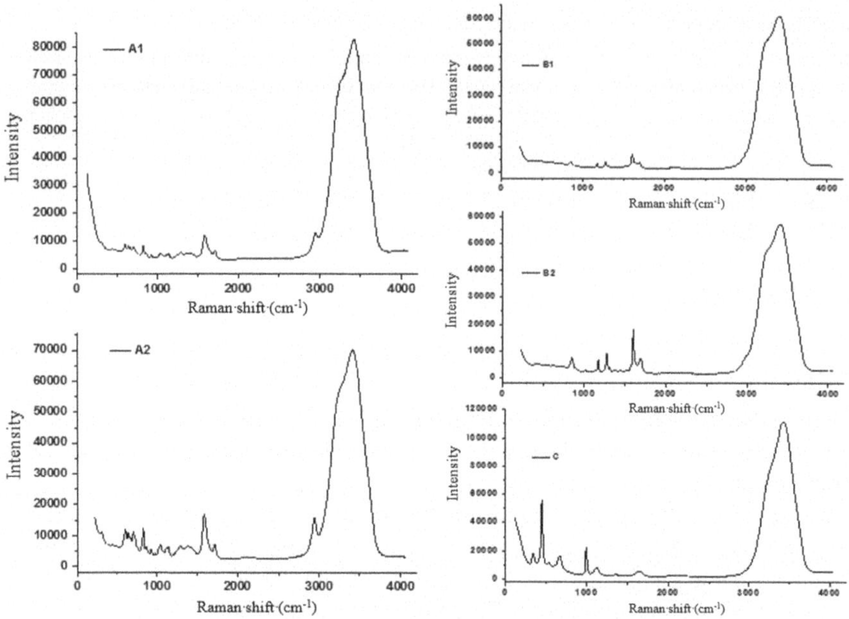

Fig. 4 Raman spectra for ascorbic acid 50 mg/cm^3 (A1), ascorbic acid 100 mg/cm^3 (A2), novocaine 5 mg/cm^3 (B1), novocaine 20 mg/cm^3 (B2) and sodium thiosulphate 300 mg/cm^3 (C)

Table 2 Factor variation limits of a multi-factor experiment

Influence factor number	Influence factor	Name of medicine	X0	Value of variation	X+	X−
1	Active ingredient content	Novocaine	5 mg/cm^3	1 mg/cm^3	6 mg/cm^3	4 mg/cm^3
			20 mg/cm^3	4 mg/cm^3	24 mg/cm^3	16 mg/cm^3
		Ascorbic acid	50 mg/cm^3	10 mg/cm^3	60 mg/cm^3	40 mg/cm^3
			100 mg/cm^3	20 mg/cm^3	120 mg/cm^3	80 mg/cm^3
		Sodium thiosulphate	300 mg/cm^3	60 mg/cm^3	360 mg/cm^3	240 mg/cm^3
2	Packaging option	For all medicines	3 cm^3	1 cm^3	4 cm^3	2 cm^3
3	Producer		0	1	Producer 2 (1)	Producer 3 (−1)
4	Temperature		24 °C	3 °C	27 °C	21 °C

Table 3 Multi-factor experiment design

Experiment number	Influence factor number			
	1	2	3	4
	Active ingredient content, *C*	Packaging option, *V*	Producer, *P*	Temperature, *T*
1	X+	X+	X+	X+
2	X−	X−	X+	X+
3	X+	X−	X+	X+
4	X−	X+	X+	X+
5	X+	X+	X−	X+
6	X−	X−	X−	X+
7	X+	X−	X−	X+
8	X−	X+	X−	X+
9	X+	X+	X+	X−
10	X−	X−	X+	X−
11	X+	X−	X+	X−
12	X−	X+	X+	X−
13	X+	X+	X−	X−
14	X−	X−	X−	X−
15	X+	X−	X−	X−
16	X−	X+	X−	X−

– for 20 mg/cm^3 novocaine content

$$\Delta\Pi = 1.010(C - C_0) - 0.005(V - V_0) - 1.320(P - P_0) + 0.092(T - T_0); \quad (3)$$

– for 50 mg/cm^3 ascorbic acid content

$$\Delta\Pi = 0.200(C - C_0) - 4.765(V - V_0) - 0.0001(P - P_0) + 1.083(T - T_0); \quad (4)$$

– for 100 mg/cm^3 ascorbic acid content

$$\Delta\Pi = 0.186(C - C_0) - 9.446(V - V_0) - 0.0001(P - P_0) + 2.172(T - T_0); \quad (5)$$

– for 300 mg/cm^3 sodium thiosulphate content

$$\Delta\Pi = 0.161(C - C_0) - 2.995(V - V_0) - 0.0005(P - P_0) + 3.588(T - T_0). \quad (6)$$

The impact of the methodological parameters was calculated using the following formulas:

- for 5 mg/cm^3 novocaine content

$$\sum_{\xi=1}^{P}\left(\frac{\partial C_i}{\partial \Delta \Pi_\xi}\right)^2 \cdot \Delta^2(\Delta \Pi_\xi) = \sqrt{\left(0.253\frac{\Delta C}{\sqrt{3}}\right)^2 + \left(0.426\frac{\Delta V}{\sqrt{3}}\right)^2 + \left(0.0001\frac{\Delta P}{\sqrt{3}}\right)^2 + \left(0.108\frac{\Delta T}{\sqrt{3}}\right)^2}; \quad (7)$$

- for 20 mg/cm^3 novocaine content

$$\sum_{\xi=1}^{P}\left(\frac{\partial C_i}{\partial \Delta \Pi_\xi}\right)^2 \cdot \Delta^2(\Delta \Pi_\xi) = \sqrt{\left(1.010\frac{\Delta C}{\sqrt{3}}\right)^2 + \left(0.005\frac{\Delta V}{\sqrt{3}}\right)^2 + \left(1.320\frac{\Delta P}{\sqrt{3}}\right)^2 + \left(0.092\frac{\Delta T}{\sqrt{3}}\right)^2}; \quad (8)$$

- for 50 mg/cm^3 ascorbic acid content

$$\sum_{\xi=1}^{P}\left(\frac{\partial C_i}{\partial \Delta \Pi_\xi}\right)^2 \cdot \Delta^2(\Delta \Pi_\xi) = \sqrt{\left(0.200\frac{\Delta C}{\sqrt{3}}\right)^2 + \left(4.465\frac{\Delta V}{\sqrt{3}}\right)^2 + \left(0.0001\frac{\Delta P}{\sqrt{3}}\right)^2 + \left(1.083\frac{\Delta T}{\sqrt{3}}\right)^2}; \quad (9)$$

- for 100 mg/cm^3 ascorbic acid content

$$\sum_{\xi=1}^{P}\left(\frac{\partial C_i}{\partial \Delta \Pi_\xi}\right)^2 \cdot \Delta^2(\Delta \Pi_\xi) = \sqrt{\left(0.186\frac{\Delta C}{\sqrt{3}}\right)^2 + \left(9.45 + \frac{\Delta V}{\sqrt{3}}\right)^2 + \left(0.0001\frac{\Delta P}{\sqrt{3}}\right)^2 + \left(2.172\frac{\Delta T}{\sqrt{3}}\right)^2}; \quad (10)$$

- for 300 mg/cm^3 sodium thiosulphate content

Table 4 Measurement results for the active ingredient content obtained using GVET 176-1-2010

Name of medicine	Active ingredient content, mg/cm^3	Measurement accuracy (absolute error bounds with the probability of $P = 0.95$), $\pm \Delta_{GVET}$, %
Ascorbic acid	50.9	2.3
	99.5	4.0
Novocaine	4.89	0.20
	21.47	0.60
Sodium thiosulphate	306.3	4.9

$$\sum_{\xi=1}^{P}\left(\frac{\partial C_i}{\partial \Delta \Pi_\xi}\right)^2 \cdot \Delta^2(\Delta \Pi_\xi) = \sqrt{\left(0.161\frac{\Delta C}{\sqrt{3}}\right)^2 + \left(2.995 + \frac{\Delta V}{\sqrt{3}}\right)^2 + \left(0.0005\frac{\Delta P}{\sqrt{3}}\right)^2 + \left(3.558\frac{\Delta T}{\sqrt{3}}\right)^2}; \tag{11}$$

where

ΔC is the active ingredient, mg/cm^3;
ΔV is the packaging option (the volume of an ampoule), cm^3;
ΔP is the pharmaceutical manufacturer;
ΔT is ambient temperature.

The measurement results for the active ingredient content obtained using GVET 176-1-2010 are given in Table 4.

The significance of the bias between the measurement results obtained using the above-mentioned procedure and those obtained using GVET 176-1-2010 was tested in accordance with [16] with the help of a t-test. The bias was established to be insignificant.

Due to the insignificance of the bias, the measurement trueness was calculated using the following formula:

$$\Delta_{c,m} = 1.96\sqrt{\frac{S_m^2}{L} + \frac{\Delta_{GVET}^2}{3} + \sum_{\xi=1}^{P}\left(\frac{\partial C_i}{\partial \Delta \Pi_\xi}\right)^2 \cdot \Delta^2(\Delta \Pi_\xi)} = 1.96 \cdot \sigma_{c,m}, \tag{12}$$

where S_m^2 is the dispersion characterizing the variation of measurement results under reproducibility conditions, mg/cm^3;

L is the number of measurement results;
Δ_{GVET} is the error of measurement results obtained using GVET 176-1-2010, mg/cm^3;

Table 5 Measurement range, accuracy, measurement repeatability and measurement reproducibility

Measurement range, mg/cm^3	Measurement repeatability (relative standard deviation of a single measurement under a set of repeatability conditions), σ_r, %	Measurement reproducibility (relative standard deviation of a single measurement under a set of reproducibility conditions), σ_R, %	Measurement accuracy (relative error bounds with the probability of $P = 0.95$), $\pm\delta$, %
From 1 to 350	5	8	20

The obtained accuracy indicators of the measurement procedure are given in Table 5.

3 Organisation and Implementation of Interlaboratory Comparisons

The next step in ensuring the quality of pharmaceutical products is the participation of testing laboratories in IC procedures in order to confirm the proper introduction of the measurement procedure into laboratory practice [21].

In 2018, UNIIM conducted special ICs for Roszdravnadzor laboratories in the field of the quantitative determination of the active ingredients of pharmaceutical products and in the field of their authentication/identification using the RS method.

Three injectable pharmaceutical products were used as test subjects in this comparison. Their active ingredients are as follows:

- 2,3-Dimercapto-1-propanesulfonic acid (synonym: 2,3-bis(sulfanyl)propane-1-sulfonic acid; unithiol) was designated as the IC sample (RS) No. 1 (see Fig. 5). The name of the pharmaceutical product is Unithiol-Ferein [22].
- 2-(2-carboxyethyl)-1,1,1-trimethyldiazanium (synonym: 3-(2,2,2-trimethylhydrazinium)propionate dihydrate, meldonium) was designated as the IC sample (RS) No. 2 (see Fig. 6) The name of the pharmaceutical substance is meldonium [23].

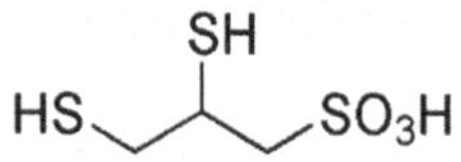

Fig. 5 Sample (RS) No. 1 for conducting the IC

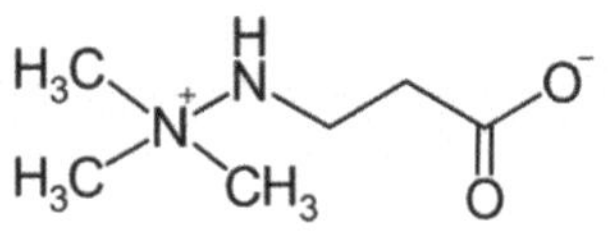

Fig. 6 Sample (RS) No. 2 for conducting the IC

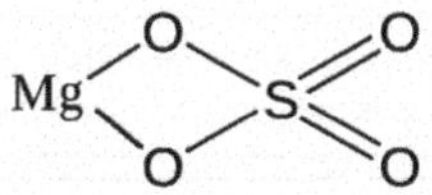

Fig. 7 Sample (RS) No. 3 for conducting the IC

Table 6 Assigned values and standard deviation for professional assessment for the active ingredient content

IC sample	Assigned values established using GVET 176-1-2010, C, mg/cm^3	σ_{r0}, %	σ_{R0}, %	σ_{L0}, %	$\hat{\sigma}_0$, %	$\hat{\sigma}$, mg/cm^3
Unithiol-Ferein (RS) No. 1	49.8	5	8	6.2	7.2	3.6
Meldonium (RS) No. 2	100.1	5	8	6.2	7.2	7.2
Magnesium sulphate (RS) No. 3	264.7	5	8	6.2	7.2	19.0

- Magnesium sulphate designated as the IC sample (RS) No. 3 (see Fig. 7). The name of the chemical compound is magnesium sulphate [24].

The pharmaceutical products in question were each produced in the volume of one batch. They did not undergo additional preparation.

The values obtained using GVET 176-1-2010 State Secondary Measurement Standard for units of mass fraction, mass (molar) concentration of components in solid and liquid substances and materials based on volumetric titration are used as the assigned values of the characteristics under study (Table 6). The metrological traceability chain is shown in Fig. 8.

The homogeneity and stability of the assigned values were studied taking into account [25]. The inhomogeneity was evaluated using the one-way variance analysis (ANOVA). For this purpose, six samples were selected. Subsequently, five measurements were performed in-parallel using GVET 176-1-2010. The results are given in Table 7.

As can be seen in Table 8, the homogeneity of the samples meets the requirements of the IC.

The standard deviation due to long-term instability was evaluated via a classical ageing method taking [24] into account.

Six samples were selected for conducting the IC. Measurements were carried out at regular intervals; the existence of a linear data trend over time was then estimated by linear approximation of the obtained data.

The measurement results are shown in Figs. 9, 10 and 11. The results of the standard deviation due to instability are presented in Table 9.

As can be seen from Table 9, the condition $|b_1| < t_{0,05;(n-2)} \cdot s(b_1)$ is met for the samples, which indicates the insignificance of the slopes of the linear approximations

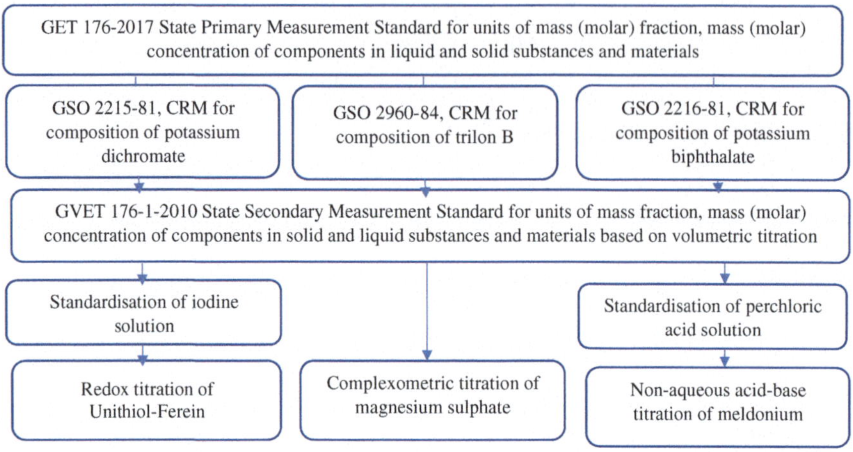

Fig. 8 Metrological traceability chain up to the GET 176-2017 state primary measurement standard

Table 7 Results of standard deviation due to inhomogeneity calculations

IC sample	Sum of squared deviations within a single sample, MS_{within}, $(mg/cm^3)^2$	Sum of squared deviations between samples, MS_{among}, $(mg/cm^3)^2$	Standard deviation due to inhomogeneity, (σ_h), mg/cm^3
Unithiol-Ferein (RS) No. 1	9.41	11.39	0.61
Meldonium (RS) No. 2	7.85	76.43	1.73
Magnesium sulphate (RS) No. 3	11.09	95.8	4.12

Table 8 Homogeneity estimation

IC sample	σ_h	$\hat{\sigma}$	$0,3 \cdot \hat{\sigma}$	Estimation
Unithiol-Ferein (RS) No. 1	0.61	3.6	1.07	Homogeneous
Meldonium (RS) No. 2	1.73	7.2	2.2	Homogeneous
Magnesium sulphate (RS) No. 3	4.12	19	5.7	Homogeneous

and the absence of a trend of the certified value during the study of the instability of the IC samples.

The measurement result quality was evaluated in accordance with [26] and [27] by comparing the z-score values with the established control standards.

All participants of IC rounds obtained satisfactory results when implementing the RS-based measurement procedure for authentication (identification) and when determining active ingredient content in injectable medicines. Results for calculating

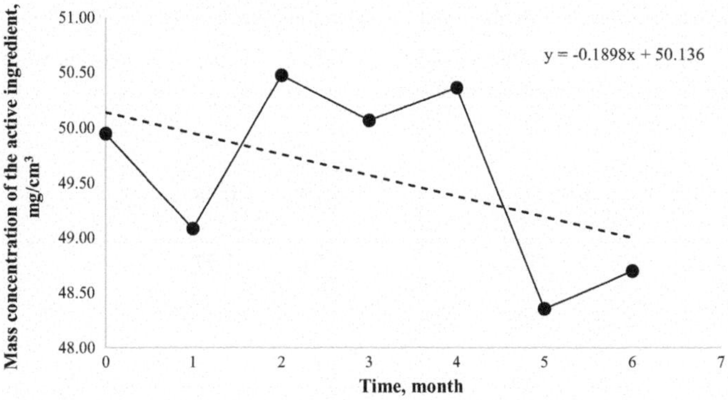

Fig. 9 Measurement results of the active ingredient content for the Unithiol-Ferein sample (RS) No. 1 when studying instability

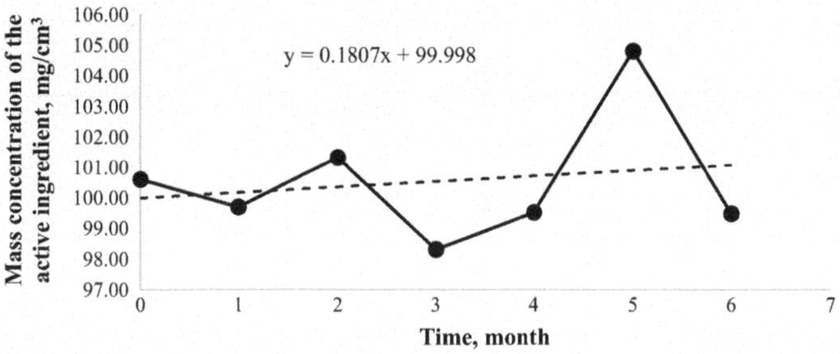

Fig. 10 Measurement results of the active ingredient content for the meldonium sample (RS) No. 2 when studying instability

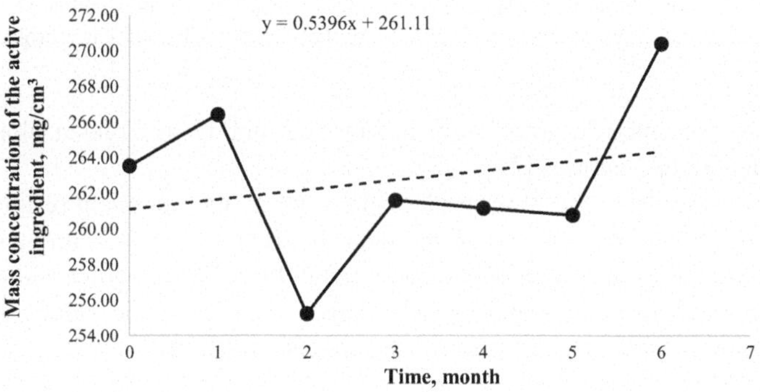

Fig. 11 Measurement results of the active ingredient content for the magnesium sulphate sample (RS) No. 3 when studying instability

Table 9 The results of evaluating the standard uncertainty due to long-term instability

IC sample	The segment cut off on the y-axis, b_0	Slope, b_1	$s(b_1)$	$t_{0,05;(n-2)} \cdot s(b_1)$	σ_{lts}
Unithiol-Ferein (RS) No. 1	50.14	−0.19	0.21	0.49	1.24
Meldonium (RS) No. 2	100.00	0.18	0.58	1.37	3.47
Magnesium sulphate (RS) No. 3	261.11	0.54	1.27	3.01	7.61

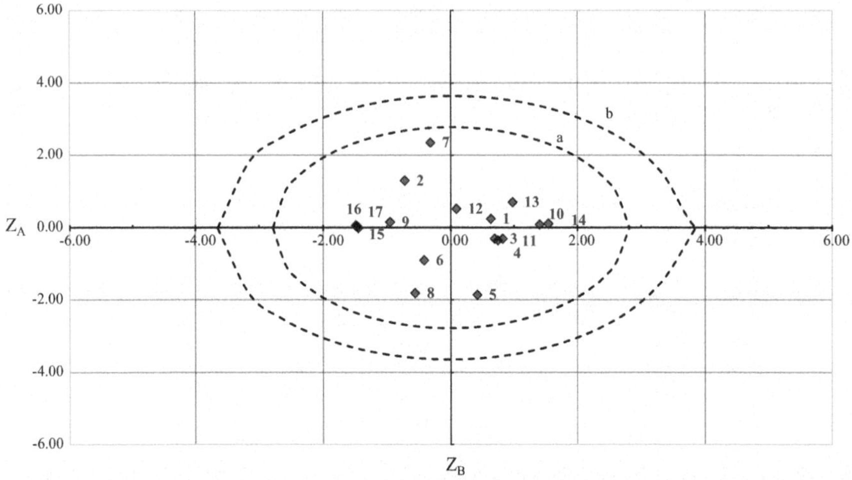

Fig. 12 Youden-plot for z-scores according to interlaboratory comparison results where Z_A denotes values of z-scores of laboratories obtained for Unithiol-Ferein; Z_B denotes values of z-scores of laboratories obtained for meldonium; **a** confidence probability 95%; **b** confidence probability 99%

z-scores are presented in graph form in Figs. 12 and 13 as Youden plots, where numbers denote laboratories [27].

As can be seen from the Youden-plots, there are no measurement results that are markedly different from the rest of the data. In addition, all measurement results are inside the ellipse having a confidence probability of 95%, which indicates that the measurement results are not biased relative to the assigned value, as well as demonstrating the adequacy of the selected analysis method.

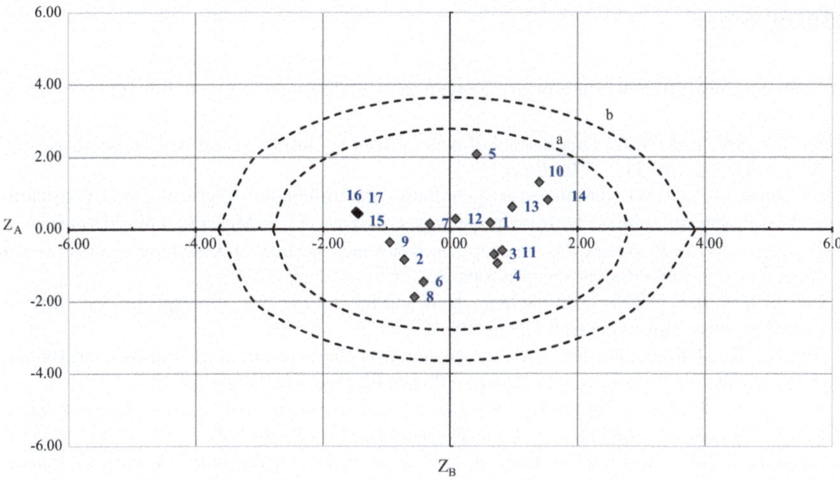

Fig. 13 Youden-plot for z-scores according to interlaboratory comparison results where Z_A are values of z-scores of laboratories obtained for Unithiol-Ferein; Z_B—values of z-scores of laboratories obtained for magnesium sulphate; **a** confidence probability of 95%; **b** confidence probability of 99%

4 Conclusions

1. An identification procedure with subsequent quantitative determination of the active ingredient of injectable medicines was certified using RS.
2. Within the framework of certifying the measurement procedure, a multi-factor experiment was conducted to assess the influence of methodological factors. The influence coefficient values were determined, which were then used to calculate the accuracy indicator of the measurement procedure.
3. ICs were set up and conducted for Roszdravnadzor laboratories in the field of medicine authentication (identification) and the determination of the active ingredient of medicines using the RS method.
4. The traceability of assigned values when carrying out interlaboratory comparisons is provided by means of performing measurements using the GVET 176-1-2010 State Secondary Measurement Standard for units of mass fraction and mass (molar) concentration of components in solid and liquid substances and materials on the basis of volumetric titration.
5. The metrological characteristics of the measurement procedure for the identification and subsequent quantitative determination of active ingredient of injectable medicines using RS were confirmed by the IC results.
6. The results of the study carried out during the IC sample study can be used to create reference materials for the active ingredient content of injectable medicines: Unithiol-Ferein, meldonium and magnesium sulphate.

References

1. Grishanov VN (2015) Raman spectroscopy in medical diagnostics. CD-ROM, SGAU, Samara (in Russ.)
2. Burikov SA et al (2009) Diagnostics of water-ethanol solutions using Raman spectroscopy. Atmos Oceanic Opt 11:1–7 (in Russ.)
3. Agrafenin AV (2014) Simultaneous quantitative determination of glycerol and potassium acetate in aqueous solution by Raman spectroscopy. Trace Elem Med 15(3):48–51 (in Russ.)
4. Tondepu C, Toth R, Navin CV, Lawson LS, Rodriguez JD (2017) Screening of unapproved drugs using portable Raman spectroscopy. Anal Chim Acta 973:75–81
5. Carron K, Cox R (2010) Qualitative analysis and the answer box: a perspective on portable Raman spectroscopy. Anal Chim Acta 82:3419–3425
6. Buckley K, Matousek P (2011) Recent advances in the application of transmission Raman spectroscopy to pharmaceutical analysis. J Pharm Biomed Anal 55:645–652
7. Strachan CJ, Rades T, Gordon KC, Rantanen J (2007) Raman spectroscopy for quantitative analysis of pharmaceutical solids. J Pharm Pharmacol 59:179–192
8. Vankeirsblick T, Vercauteren A, Baeyens W, van der Weken G, Verpoort F, Vergote G, Remon JP (2002) Applications of Raman spectroscopy in pharmaceutical analysis. TrAC Trends Anal Chem 21:869–877
9. Connie M, Gryniewicz-Ruzicka Rodriguez JD, Arzhantsev S, Buhse LF, Kauffman JF (2012) A comparison of chemometric methods for the analysis of Raman spectra of contaminated pharmaceutical materials. J Pharm Biomed Anal 61:191–198
10. Hennigan MC, Rydes AG, Gordon KC, Rantanen J (2007) Raman spectroscopy for quantitative analysis of pharmaceutical solids. J Pharm Pharmacol 59:179–192
11. Matero S, Den Berg FV, Poutiainen S, Rantanen J, Pajander J (2013) Towards better process understanding: chemometrics and multivariate measurements in manufacturing of solid dosage forms. J Pharm Sci 102:1358–1403
12. Rantanen J (2007) Process analytical application of Raman spectroscopy. J Pharm Sci 59:171–177
13. OFS.1.2.1.1.0009.15 Raman spectroscopy. A general pharmacopoeial monograph. Ministry of Health of the Russian Federation. http://pharmacopoeia.ru/wp-content/uploads/2016/10/OFS.1.2.1.1.0009.15-Ramanovskaya-spektrometriya.pdf (in Russ.)
14. Galeev R, Ma B, Saveliev A, Iishat A (2015) Quantitative analysis of injectable drug products using non-invasive wideband Raman technology. Am Pharm Rev 18(6)
15. Lemeshko BYu (2018) Criteria for testing the distribution deviation from the normal law: application guide. INFRA-M, Moscow (in Russ.)
16. ISO 5725-2:1994 Accuracy (trueness and precision) of measurement methods and results—Part 2: Basic method for the determination of repeatability and reproducibility of a standard measurement method
17. CCQM-K130 (2018) The international key comparison database. BIPM, France. https://kcdb.bipm.org/appendixB/KCDB_ApB_info.asp?cmp_idy=1472&cmp_cod=CCQM-K130&prov=exalead. Accessed 30 Oct 2018
18. CCQM-K149 International Key Comparisons (2018) BIPM, France. https://kcdb.bipm.org/appendixB/KCDB_ApB_info.asp?cmp_idy=2625&cmp_cod=CCQM-K149&prov=exalead. Accessed 30 Oct 2018
19. Medvedevskikh M et al (2017) Key comparison final report of CCQM-K130 nitrogen mass fraction measurements in glycine. Metrologia 54:08004. https://doi.org/10.1088/0026-1394/54/1A/08004
20. Derffel K (1994) Statistics in analytical chemistry. Mir, Moscow (in Russ.)
21. ISO/IEC 17025:2017 General requirements for the competence of testing and calibration laboratories
22. PN002691/01 Manufacturer's pharmacopoeial monograph. Solution for intramuscular and subcutaneous injection 5%. Unithiol-Ferein (in Russ.)

23. LSR-000864/10-100210 The quality standard of pharmaceutical products. Manufacturer's pharmacopoeial monograph. Idrinol, a solution for intravenous and parabulbar injection of 100 mg/ml. Meldonium (in Russ.)
24. ND 8570-03 Magnesium sulphate. Solution for intravenous and intramuscular injection 200 and 250 mg/ml (in Russ.)
25. ISO GUIDE 35:2017 Reference materials—Guidance for characterization and assessment of homogeneity and stability
26. ISO/IEC 17043:2010 Conformity assessment—General requirements for proficiency testing
27. ISO 13528:2015 Statistical methods for use in proficiency testing by interlaboratory comparison

Reference Materials for Isotopic Composition Used in Implementing Isotope Dilution Mass Spectrometry

Olga N. Kremleva and Tatiana N. Tabatchikova

Abstract The article presents information on the development of reference materials for enriched isotopic composition. The possibility of using reference materials for isotopic composition in the implementation of isotope dilution mass spectrometry is considered. The authors show the capabilities of inductively coupled plasma mass spectrometry in measuring the isotopic composition of various elements.

Keywords Mass spectrometry · Inductively coupled plasma mass spectrometry · Isotopic dilution mass spectrometry · Certified reference material · GSO CRM registration number in the National Register of the Russian Federation

1 Introduction

Depending on the field of application, mass spectrometry (MS) acts as the main or one of the main instrumental methods of chemical or biochemical analysis. One variation of the method, i.e. isotope dilution mass spectrometry (IDMS), is recognized by the Consultative Committee for Amount of Substance (CCQM) as a method, which can be considered a primary measurement method [1]. It is hard to imagine an area of human activity where mass spectrometry is not used. We shall confine ourselves to simply listing the following areas: biochemistry, clinical chemistry, general chemistry, organic chemistry, pharmaceutics, cosmetics, perfumery, food industry, chemical synthesis, petrochemistry, oil refining, environmental monitoring, production of polymers and plastics, medicine and criminology, doping control, control of narcotic drugs, alcohol control, geochemistry, geology, hydrology, petrography, mineralogy, geochronology, archaeology, nuclear and energy industries, semiconductor industry, metallurgy.

Main trends in the development of modern MS can be analysed using scientific (bibliometric) indicators, such as the number of publications and references to publications, and economic indicators, such as the market volume of various mass spectrometers. These indicators are quite fully covered in [2]. This our work aims

O. N. Kremleva (✉) · T. N. Tabatchikova
Ural Scientific Research Institute for Metrology (UNIIM), Ekaterinburg, Russia
e-mail: kremleva@uniim.ru

S. Medvedevskikh et al. (eds.), *Reference Materials in Measurement and Technology*,
https://doi.org/10.1007/978-3-030-32534-3_22

to continue analytical studies on the development of MS techniques using scientific indicators.

The isotopic dilution technique in MS has a number of advantages over other analytical techniques. However, despite the advantages, it has some significant limitations [3]. The main problem when using this technique consists in the insufficient range of characterised materials of isotopically-enriched pure substances. Isotopically-enriched pure substances are rather expensive, especially those, in which ratios are strongly altered relative to the natural mixture of isotopes (each fraction of atomic percent increases the cost of the substance) [3].

In the past, MS methods were not broadly applied, largely due to an insufficient provision of appropriate MS equipment. However, the situation has significantly changed for the better over the past decades. According to the available data of a large marketing company Techart.Research [4], the market of mass spectrometers had shown a significant growth by the end of 2010, amounting to about $2.6 billion. Similar growth rates are expected for 2011.

While being currently equipped with necessary instrumentation, MS (particularly IDMS) often lacks not only sufficient methodological and metrological support, but also highly qualified human resources outside of scientific institutions. As a result, the isotope dilution technique is not widespread, with a rather small group of reference laboratories in different countries apply it as a referee method.

Thus, the primary task in developing the isotope dilution technique of inductively coupled plasma mass spectrometry (ICP-MS) consists in providing methodological support, including the development and creation of a range of pure substances (reference materials—RMs) for isotopic composition of elements. By 2011, the State Register of type-approved (certified) RMs of the Russian Federation had contained only 3 isotopically-enriched certified reference materials (CRM): CRM for isotopic composition of thallium (III) oxide, enriched with thallium-205 isotope (GSO 7550-99), CRM for isotopic composition of calcium carbonate, enriched with calcium-48 isotope (GSO 7701-99), CRM for composition of boric acid, enriched with boron-10 isotope (GSO 8544-2004). Since 2012, the Scientific Methodological Centre of State Service Department[1] of the Ural Scientific Research Institute for Metrology (UNIIM) has been studying the possibility of using IDMS for creating reference materials of substances and materials, simultaneously developing CRMs for isotopic composition. The studies are carried out using a PerkinElmer NexION 300D ICP Mass Spectrometer, included in the State Secondary Measurement Standard for the Mass Fraction and Mass (Molar) Concentration of Metals in Solid and Liquid Substances and Materials GVET 196-1-2012.

To date, the following range of RMs have been developed:

- CRM for the isotopic composition of nickel, enriched by ^{60}Ni isotope in nitrate solution (^{60}Ni CRM UNIIM) —GSO 10273-2013;
- CRM for isotopic composition of lead, enriched by ^{206}Pb isotope, in nitrate solution (^{206}Pb CRM UNIIM)—GSO 10274-2013;

[1] State Service of Reference Materials for Composition and Properties of Substances and Materials.

- CRM for isotopic composition of cadmium, enriched by ^{111}Cd isotope, in nitrate solution (^{111}Cd CRM UNIIM)—GSO 10493-2014;
- CRM for isotopic composition of silver, enriched by ^{107}Ag isotope, in nitrate solution (^{107}Ag CRM UNIIM)—GSO 10494-2014.

Considering a strong need for such CRMs, RM materials for enriched isotopic composition of chromium, thallium, magnesium, strontium, rubidium, copper, zinc, etc., have been prepared for further development.

Along with the development of the above-mentioned CRMs, UNIIM is testing and implementing the IDMS method using RMs developed in the course of various measurements pertaining to scientific work of the State Service Department, as well as during RM testing for the purposes of type approval. Three research directions of this project have become the subject of the present research.

2 Scientific Indicators of the Development of the Method

The search for scientific articles on MS were carried out using two databases:

- scientific electronic library eLIBRARY.RU, the largest Russian information portal in the field of science, technology, medicine and education, containing abstracts and full texts of more than 14 million scientific articles and publications [5];
- ScienceDirect, one of the leading full-text scientific databases offering journal articles and book chapters from more than 25,000 journals and nearly 20,000 books [6].

2.1 eLIBRARY.RU Search Results

For the primary search through the scientific electronic library eLIBRARY.RU, the following search terms were used: mass spectrometry, mass spectrometer as well as some related terms and their abbreviations (in publication titles, abstracts, keywords). The search was conducted on all types of publications included in the base, such as journal articles, books, conference materials, deposited manuscripts, theses, reports and patents. The proportions of MS development in Russia, characteristic of the last twenty years, can be seen in Fig. 1.

A narrow search was carried out using the following terms: isotope dilution mass spectrometry, isotope dilution as well as some related terms and their abbreviations (in publication titles, abstracts, keywords). The search system did not provide any works with the specified parameters published until 2004. The proportions of IDMS development in Russia, characteristic of the last twenty years, are presented in Fig. 2.

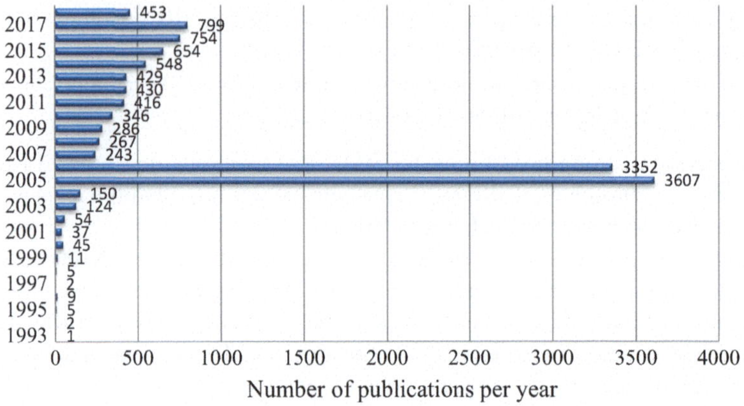

Fig. 1 Number of publications on MS for the period from 1993 to 2018 (as of October 2018), provided by the scientific electronic library eLIBRARY.RU [5]

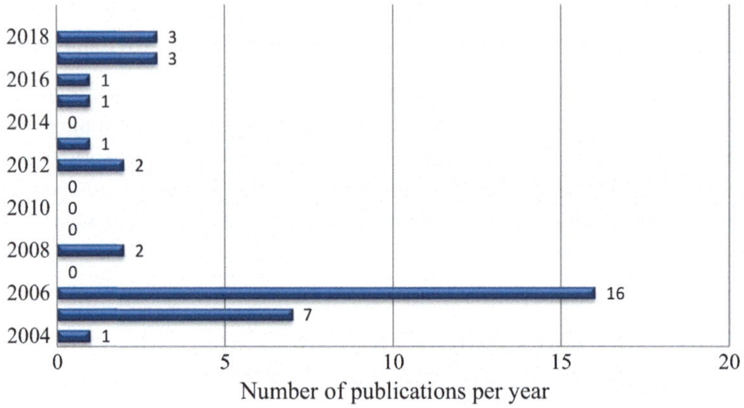

Fig. 2 Number of publications on IDMS for the period from 2004 to 2018 (as of October 2018), provided by the scientific electronic library eLIBRARY.RU [5]

2.2 *ScienceDirect Search Results*

For the primary search through the scientific database ScienceDirect, the following search terms were used: mass spectrometry, mass spectrometer as well as some related terms and their abbreviations (in publication titles, abstracts, keywords). The search was conducted on all types of publications included in the base, such as journal articles, books, conference materials, deposited manuscripts, theses, reports and patents. The total number of discovered articles, publications, etc., amounts to more than 500,000. The proportions of MS development abroad, characteristic of the last twenty years, are shown in Fig. 3.

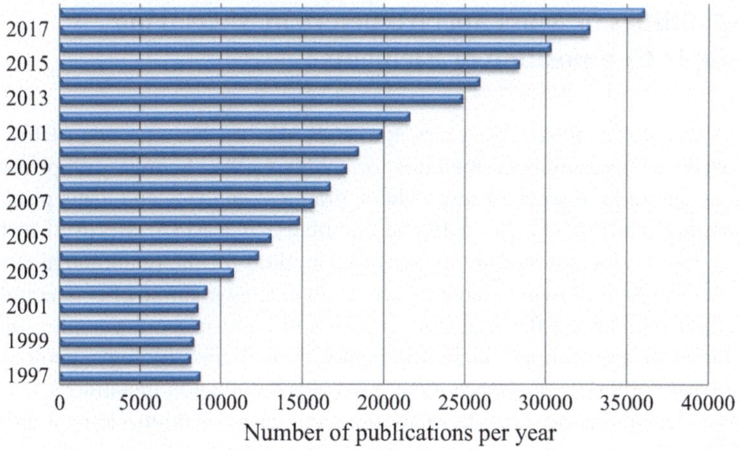

Fig. 3 Number of publications on MS for the period from 1997 to 2018 (as of October 2018), provided by the international scientific database ScienceDirect [6]

A narrow search using the term *isotope dilution mass spectrometry* showed much fewer publications (Fig. 4) as compared to the number of publications on general MS.

The analysis of publications shows an ongoing scientific interest in MS. Despite the general interest, a clear lack of publications on the subject of IDMS is noted, which is possibly due to an insufficient application of this technique.

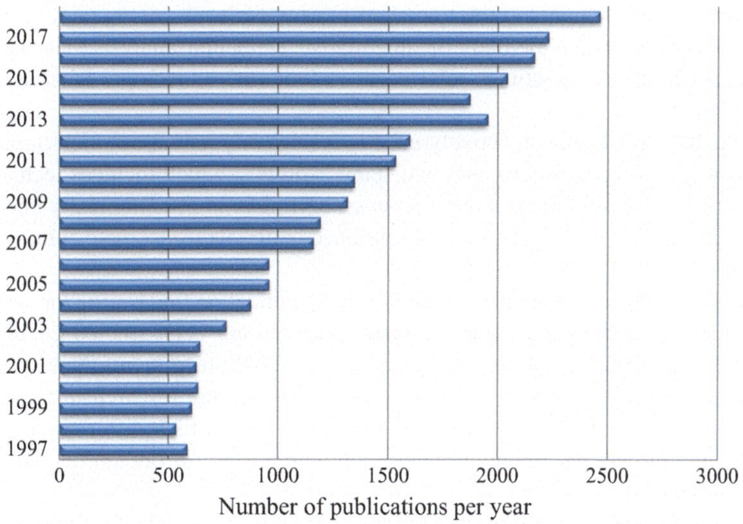

Fig. 4 Number of publications on IDMS for the period from 1997 to 2018 (as of October 2018), provided by the international scientific database ScienceDirect [6]

3 Capabilities of Mass Spectrometers in Measuring Isotopic Composition of Elements

Since its appearance thirty-five years ago, ICP-MS has become one of the most widely used and best-equipped methods for analysing the chemical composition of substances due to hard work of researchers, practical analysts and firms producing corresponding instruments. In analytical chemistry, ICP-MS has proved to be an extremely rapid, efficient and highly sensitive method for the quantitative simultaneous determination of many elements across an extensive range of concentrations. This method is intended for analysing liquid, solid and gaseous samples and has a wide range of applications, including ecology; medicine; biology; geology and geochemistry; forensics; pharmaceutical, food, semiconductor, metallurgical, chemical and nuclear industries, etc. In many practical cases of multi-element analysis, ICP-MS has provided serious competition to the methods of atomic spectroscopy traditionally used for this purpose or even completely replaced them [3].

Measurement of isotope ratios is performed in a wide range of cases: in determining the content of stable isotopes and long-lived radionuclides for studying isotope variations in the nature and as part of geochronological studies; in the quality control of nuclear fuel and in monitoring radioactive and other environmental pollution; in detecting falsified products and monitoring their quality, etc. [3].

In addition, measurement of isotopic ratios is of great importance in biological and medical studies that employ labelling with enriched stable or unstable (but relatively long-lived) isotopes [7–12]; in the study of metabolism and chemical reactions; as well as in using the isotope dilution technique (ID) to determine the content of elements in different samples. This technique takes priority when determining the concentrations of elements at the trace and ultra-trace levels, as its high accuracy is unattainable by other methods of quantitative elemental analysis. The isotope dilution technique is sometimes referred to in literature as isotopic labelling or a tracer method.

Lately, ICP-MS has been actively used in a relatively new area—determination of the isotopic composition of enriched stable isotopes, which includes technological control of the isotope separation process, certification of isotopically-enriched products, as well as characterisation of reference materials for isotopic composition [13–24].

However, we believe that this area of ICP-MS application (measuring the isotopic composition of stable highly enriched isotopes) has not been studied enough by researchers and firms developing and producing ICP-MS instruments. Nevertheless, this area is of great practical importance and is of definite research interest both in terms of ranges of measured isotopic concentrations, which differ significantly from the natural isotopic abundances, as well as methods for obtaining qualitative analytical data [3].

With certain instrumentation and well-developed methodological techniques, isotopic mass spectrometry with inductively coupled plasma is a powerful method for performing correct and reproducible isotopic analysis. In recent years, this method has become a serious competitor to thermal ionisation mass spectrometry (TIMS) in this analytical field. Until the end of the 20th century, TIMS was generally accepted and standard for high-precision isotopic analysis. For example, most of IUPAC recommendations (International Union of Pure and Applied Chemistry) on establishing exact atomic weights of elements [25] are based on measuring their natural isotope abundance using TIMS. This method has a much longer history than ICP-MS, a huge bibliography (see, for example, a section on TIMS in [26]) and is based on evaporating the studied element from the matrix of a condensed sample and thermal ionization of its atoms to produce a highly stable ion beam. For routine studies (for example, determining the age of geological rocks), MS has traditionally employed instruments of multi-collector thermal ionisation mass spectrometry (MC-TIMS) to attain isotopic ratios with an accuracy of less than 0.01%.

A growing role of ICP-MS in isotopic analysis is evidenced by a rather rapid increase in the number of publications devoted to this area of analytical chemistry. Various types of instruments are used for this purpose, such as quadrupole (ICP-Q-MS), time-of-flight (ICP-TOF-MS), magnetic sector-field (ICP-SF-MS), as well as high resolution instruments (ICP-HR-MS) with double focusing in magnetic sector and electric fields (ICP-DF-MS) having a single-collector (ICP-DF-MS-SC) and, especially, multi-collector (ICP-DF-MS-MC) system of ionic current registration. Isotopic analysis is performed using instruments of various manufacturers and even laboratory-made instruments.

The use of quadrupole mass spectrometers for rapid multi-element analysis has several advantages: rapid sample entry at normal atmospheric pressure and shorter measurement time (1 min). With the same accuracy of measuring isotopic ratios, the level of uncertainty is rather low (about 0.1%).

In [28], the authors compared the data on establishing strontium isotope ratios (^{87}Sr/^{86}Sr) in SRM 987 Strontium Carbonate Isotopic Standard (NIST, USA) obtained in different years using different types of mass spectrometers.

Having the above-mentioned reference material, UNIIM joined this endeavour and presented additional data on the possibilities of using an ICP-Q-MS (NexION 300D, Perkin Elmer, USA) to determine isotopic ratios. The summarised data is given in Table 1.

All results presented in Table 1 imply performing measurements of 10 samples. A comparison of the obtained results shows that the use of "fast" quadrupole mass spectrometers instead of "slow" high-precision instruments having high resolution to establish strontium isotopic ratios significantly reduces the time required for sample preparation and measurements. The use of quadrupole mass spectrometers proposed in [27] implies a time expenditure of 100 min for measuring 10 samples versus 720 min for measuring the same samples using TIMS instruments.

The results obtained in a UNIIM laboratory in 2014 are in good agreement with the data provided by various foreign laboratories. Thus, it can be assumed that a reference material for isotopic composition of strontium enriched with ^{86}Sr, whose metrological characteristics will be established using a NexION 300D quadrupole mass spectrometer, will not be inferior to analogous reference materials for isotopic composition in terms of accuracy indicators.

Table 1 Comparison of measured and certified values of establishing strontium isotope ratios in SRM 987 Strontium Carbonate Isotopic Standard (NIST, USA)

Method (instrument type)	Certified value	$^{87}Sr/^{86}Sr$	±RSD, %	Year of publication [27]
TIMS	$^{87}Sr/^{86}Sr =$ 0.71026(12)	0.71023	0.00245	1985
ICP-MS-MC		0.70316	0.0025	2001
ICP-SF-MS-MC		0.71020	0.0001	2002
ICP-SF-MS		0.7044	0.2	2001
ICP-LA-MS		0.7103	0.084	2002
ICP-Q-MS		0.721	0.41	2001
ICP-SF-MS		0.71029	0.057	2003
ICP-Q-MS		0.71090	0.22	2003
ICP-Q-MS		**0.7101**	**0.0058**	**2014, UNIIM**

4 IDMS Potential in Determining Certified Values of the CRM for the Composition of Formula Milk

In 2011, a CRM for composition of formula milk (SM CRM UNIIM) registered under GSO 9968-2011 was developed by efforts of State Service Department (UNIIM). Detailed information on the development of this reference material is presented in [28], whereas its metrological characteristics are presented in Table 2.

In the course of type-approval testing of the reference material in question, mass fractions of elements (copper, cadmium, arsenic, lead and mercury) were measured via ICP-MS employing the calibration curve technique. The mass fraction of lead determined using the above-mentioned technique amounted to (0.028 ± 0.006) ppm (μg/g). The relative error bounds of mercury mass fraction exceed 20%; thus, according to the "Specification for developing certified reference material for composition of formula milk (SM CRM UNIIM)", mercury mass fraction cannot be included in certified characteristics. In the CRM passport, it is indicated only as an additionally established characteristic.

In order to compare the accuracy of results obtained measuring the mass fraction of lead in formula milk via IDMS and ICP-MS employing the calibration curve technique, the mass fraction of lead in the CRM for composition of skimmed dry milk

Table 2 Metrological characteristics of GSO 9968-2011 CRM for the composition of formula milk (batch 2)

Certified characteristic	Element	Certified value	Relative error bounds (at P = 0.95), δ, %
Mass fraction of elements, ppm (μg/g)	Copper	4.7	±8.5
	Cadmium	0.026	±15.4
	Arsenic	0.088	±14.8

Table 3 Metrological characteristics of GSO 9836-2011 CRM for composition of skimmed dry milk (CRM BCR-151)

Certified characteristic	Element	Certified value	Expanded uncertainty of certified value (at k = 2), *U*, ppm
Mass fraction of elements, ppm (μg/g)	Cadmium	0.10	0.008
	Copper	5.23	0.08
	Iron	50.1	1.3
	Mercury	0.101	0.010
	Iodine	5.35	0.14
	Lead	2.002	0.026

(CRM BCR-151) GSO 9836-2011, as well as in the above-mentioned GSO 9968-2011 was established using these methods. Table 3 presented below contains information on the metrological characteristics of GSO 9836-2011 CRM for composition of skimmed dry milk.

A lead solution having lead mass fraction of 3.68 μg/g was used as an additive solution when measuring the mass fraction of lead in the indicated reference materials via IDMS. The solution was prepared by diluting the reference material for isotopic composition of lead, enriched by ^{206}Pb isotope, in nitrate solution (^{206}Pb CRM UNIIM) GSO 10274-2013. The metrological characteristics of GSO 10274-2013 are presented in Table 4. In order to measure the mass fraction of lead using IDMS, the same amount of the above-mentioned additive solution was introduced into all of the analysed samples.

The results of measuring lead mass fraction in the GSO 9836-2011 CRM for composition of skimmed dry milk (CRM BCR-151) and GSO 9968-2011 CRM for composition of formula milk using the techniques in question are given in Tables 5 and 6.

In Figs. 5 and 6, the aforementioned results of measuring the mass fraction of lead are presented graphically.

The mean of the results of the lead mass fraction measurements ($\bar{X}$) for the indicated reference materials, obtained using different mass spectrometry techniques,

Table 4 Metrological characteristics of GSO 10274-2013 CRM for isotopic composition of lead, enriched by ^{206}Pb isotope, in nitrate solution (^{206}Pb CRM UNIIM)

Certified characteristic		Certified value	Absolute error bounds (at P = 0.95), Δ, %
Atomic fraction of Pb isotopes, %	^{204}Pb	0.0254	±0.0005
	^{206}Pb	96.082	±0.014
	^{207}Pb	2.171	±0.006
	^{208}Pb	1.722	±0.010

Table 5 Results of measuring the mass fraction of lead in the GSO 9836-2011 CRM for the composition of skimmed dry milk (CRM BCR-151) via IDMS and ICP-MS employing the calibration curve technique

Characteristic	Measurement results obtained using various techniques, μg/g	
	Calibration curve technique	Isotope dilution technique
Mass fraction of lead, μg/g	1.969	2.022
	1.959	2.000
	2.038	2.022
	1.977	2.002
	1.994	2.008
	1.968	2.015
	1.958	2.008
	2.033	1.995
	2.049	2.016
	2.031	2.008

Table 6 Results of measuring the mass fraction of lead in the GSO 9968-2011 CRM for the composition of formula milk via IDMS and ICP-MS employing the calibration curve technique

Characteristic	Measurement results obtained using various techniques, μg/g	
	Calibration curve technique	Isotope dilution technique
Mass fraction of lead, μg/g	0.0279	0.0290
	0.0278	0.0287
	0.0299	0.0290
	0.0280	0.0283
	0.0271	0.0286
	0.0274	0.0287
	0.0298	0.0281
	0.0289	0.0285
	0.0277	0.0284
	0.0298	0.0286

as well as repeatability standard deviations (S) at $N = 10$ measurement results are given in Table 7.

As the error associated with plotting a calibration curve constitutes a significant part of the error budget, which is not the case when using the isotopic dilution technique, we can draw the following conclusion. The use of IDMS provides well-reproducible and comparable values of lead mass fraction in the analysed reference materials, thus giving more accurate measurement results. In this connection, IDMS may be more preferable than ICP-MS employing the calibration curve technique when determining the metrological characteristics of reference materials for the composition of formula milk.

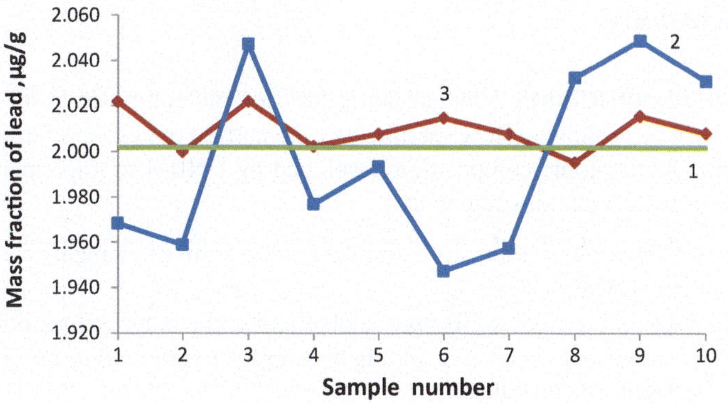

Fig. 5 Results of measuring the mass fraction of lead in the GSO 9836-2011 CRM for composition of skimmed dry milk (CRM BCR-151) using different techniques of mass spectrometry: 1—certified value of lead mass fraction in GSO 9836-2011 CRM for composition of skimmed dry milk (CRM BCR-151); 2—ICP-MS employing the calibration curve technique; 3—IDMS

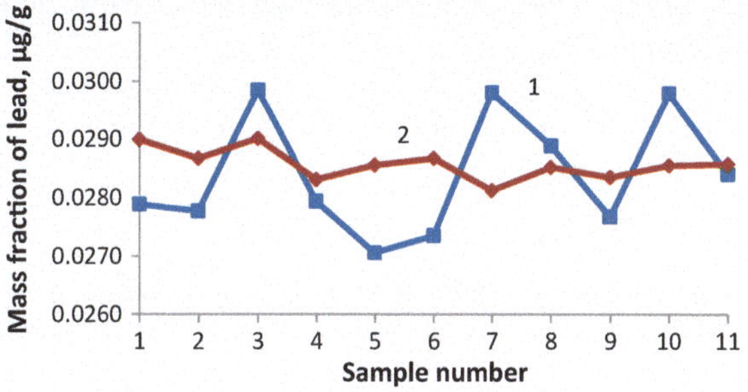

Fig. 6 Results of measuring the mass fraction of lead in the GSO 9968-2011 CRM for composition of formula milk (SM CRM UNIIM) using different techniques of mass spectrometry: 1—ICP-MS employing the calibration curve technique; 2—IDMS

Table 7 Error characteristics of the lead mass fraction measurement results obtained for the analysed CRMs using different mass spectrometry techniques

Test subject	Certified (established) value of Pb mass fraction, µg/g	Measurement results obtained via different techniques, µg/g			
		Calibration curve technique		Isotope dilution technique	
		$\bar{X}$	S	$\bar{X}$	S
CRM for composition of skimmed dry milk (CRM BCR-151) GSO 9836-2011	2.002 ± 0.026	1.996	0.038	2.010	0.009
CRM for composition of formula milk (SM CRM UNIIM) GSO 9968-2011	0.028 ± 0.006	0.0284	0.0010	0.0286	0.0003

5 Conclusions

The capabilities of ICP-MS in determining the isotopic composition of reference materials have been shown. The experiments were carried out using reference materials for enriched isotopic composition developed by UNIIM to implement IDMS. The obtained results demonstrate that:

- developed CRMs for isotopically-enriched composition of elements can be used to implement IDMS;
- the use of IDMS provides well-reproducible and comparable values of the mass fractions of impurity elements, including in complex matrices, thus ensuring more accurate measurement results;
- frequent use of IDMS allows sample preparation time to be reduced.

At UNIIM, the development of reference materials is carried out by the Group for Research and Methodological Work represented by Olga Kremleva (acting head of State Service Department) and Tatiana Tabatchikova (leading engineer of laboratory 251) with the help of specialists from the State Service Department in the capacity of experts in RM testing.

References

1. Minutes from the fifth meeting (1998) of the Consultative Committee on the Quantity of Material (CCQM) of the Bureau International des Poids et Mesures (BIPM), Sèvres, France (1998)
2. Milman B, Konopelko LA (2006) Modern mass spectrometry: proportions of development. Mass Spectrom 3(4):271–276
3. Pupyshev AA, Sermyagin BA (2006) Mass discrimination of ions in isotope analysis using inductively coupled plasma-mass spectrometry. USTU, Yekaterinburg (in Russ.)
4. Marketing research of mass spectrometers. Techart.Research, 2011. Available via http://research-techart.ru/report/mass-spectrometer-market.htm. Accessed 30 Oct 2018
5. Scientific electronic library eLIBRARY.RU. Available via https://elibrary.ru/. Accessed 30 Oct 2018
6. ScienceDirect database. Available via https://www.sciencedirect.com. Accessed 12 Feb 2019
7. Becker JS, Dietze HJ (1998) Ultratrace and precise isotope analysis by double-focusing sector field inductively coupled plasma mass spectrometry. J Anal At Spectrom 13:1057–1063
8. Barnes R (1996) Analytical plasma source mass spectrometry in biomedical research. Fresenius J Anal Chem 355:433–441
9. Lambertsson L, Lundberg E, Nilsson M, Frech W (2001) Applications of enriched stable isotope tracers in combination with isotope dilution GC-ICP-MS to study mercury species transformation in sea sediments during in situ ethylation and determination. J Anal At Spectrom 16(11):1296–1301
10. Lee KH, Jiang SJ, Liu HW (1998) Determination of mercury in urine by electrothermal vaporization isotope dilution inductively coupled plasma mass. J Anal At Spectrom 13(11):1227–1231
11. Lee KH, Liu SH, Jiang SJ (1998) Determination of cadmium and lead in urine by electrothermal vaporization isotope dilution inductively coupled plasma mass spectrometry. Analyst 7:1557–1560

12. Nobrega JA, Gelinas Y, Krushevska A, Barnes RM (1997) Determination of elements in biological and botanical materials by inductively coupled plasma atomic emission and mass spectrometry after extraction with a tertiary amine reagent. J Anal At Spectrom 12(10):1239–1242
13. Sermyagin BA, Skorikov AI, Sycheva LYu, Borodin ES (2003) Isotopic analysis of isotopically-enriched substances using inductively coupled plasma mass spectrometry. In: Conference proceedings of the 16th ural conference on spectroscopy (9–12 Sept 2003, Novouralsk), Ekaterinburg, pp 121–139 (in Russ.)
14. Sermyagin BA, Kokareva LYu, Ermakova OV (2005) Analysis of the isotopic composition of isotope-enriched substances using inductively coupled plasma mass spectrometry. In: Conference proceedings of the 17th ural conference on spectroscopy (12–15 Sept 2005, Novouralsk), Ekaterinburg, pp 84–85 (in Russ.)
15. Sermyagin BA, Skorikov AI, Cherepanov LN (2002) Experience of a mass-spectrometry laboratory in the quality control of stable isotopes produced by electromagnetic method. Analytics Control 6(3):285–288 (in Russ.)
16. Kashcheev NA, Dergachev VA (1989) Electromagnetic separation of isotopes and isotope analysis. Energoatomizdat, Moscow, 166 p (in Russ.)
17. Kovalev ID, Potapov AM (2000) High-purity isotopically-enriched substances and problems of their analytical control. Partners Competitors 11:23–28 (in Russ.)
18. Kovalev ID, Potapov AM, Suchkov AI (2003) Laser mass-spectrometric isotope analysis of isotope enriched substances. In: Programme and abstracts of the scientific contributions at 2003 European winter conference on plasma spectrochemistry, Garmisch-Partenkirchen, Germany, 12–17 Jan 2003, p 357
19. Kovalev ID, Malyshev KN, Potapov AM, Sushkov AI (2001) Isotopic analysis of silicon enriched with ^{28}Si using laser mass spectrometry. J Anal Chem 56(5):437–442 (in Russ.)
20. Kokareva LYu, Sermyagin BA (2005) Analysis of the isotopic composition of enriched barium isotopes using inductively coupled plasma mass spectrometry. In: Conference proceedings of the 17th ural conference on spectroscopy (12–15 Sept 2005, Novouralsk), Ekaterinburg, pp 85–87 (in Russ.)
21. Ermakova OV, Kokareva LYu, Sermyagin BA (2005) Analysis of the isotopic composition of enriched calcium isotopes via inductively coupled plasma mass spectrometry using cold plasma and a synthetic standard for isotopic composition to correct mass bias. In: Conference proceedings of the 17th ural conference on spectroscopy (12–15 Sep 2005, Novouralsk), Ekaterinburg, pp 87–88. (in Russ.)
22. Ermakova OV, Kokareva LYu, Sermyagin BA (2005) Analysis of the isotopic composition of enriched palladium isotopes. In: Conference proceedings of the 17th ural conference on spectroscopy (12–15 Sept 2005, Novouralsk), Ekaterinburg, pp 89–90 (in Russ.)
23. Kramchaninov AJu, Chernyshev IV, Shatagin KN (2012) Isotope analysis of strontium by multi-collector inductively-coupled plasma mass spectrometry: high-precision combined measurement of $^{88}Sr/^{86}Sr$ and $^{87}Sr/^{86}Sr$ isotope ratios. Mass Spectrom 9(2):129–138 (in Russ.)
24. Saprygin AV, Elistratov OV, Kalashnikov VA, Kulik IS, Rodichev IA (2011) Mass spectrometry determination of the isotopic composition of uranium hexafluoride. Mass Spectrom 8(1):20–26 (in Russ.)
25. De Laeter JR, Bohlke JK, De Bievre P, Hidaka H, Peiser HS, Rosman KJR, Taylor PDP (2003) Atomic weights of the elements: review 2000. Pure Appl Chem 75(6):683–800
26. Bacon JR, Crain JS, Van Vaeck L, Williams JG (2001) Atomic spectrometry update. Atomic mass spectrometry. J Anal At Spectrom 16(6):879–915
27. Zoriy MV, Rashad A, Pickhardt C, Mohsen HT, Forstel H, Helal AI, Zahran NF, Becker JS (2003) Routine method for $^{87}Sr/^{86}sr$ isotope ratio measurements in biological and geological samples after trace matrix separation by ICP-MS. At Spectrosc 24(6):195–200
28. Report on work performed within the state contract No. 120-177 of 01.06.2011. "Development of high-precision reference materials of solid inorganic and organic substances (materials) on the basis of primary methods (coulometry, gravimetry, mass spectrometry)", Yekaterinburg, 2011 (in Russ.)

Combined Measurement Uncertainty for *pH*-Values Using Certified Reference Materials in Potentiometric Measurements with Glass Electrodes

Samuel Wunderli

Abstract The practical realisation of *pH*-measurements with Harned-cell type equipment used for the measurement of primary standards as recommended by IUPAC is shortly outlined. In contrast to the Harned-cell without transference, the glass electrode with a junction is preferred and more practical for field laboratories. The combined measurement uncertainty is evaluated step by step in a systematic way according to ISO-GUM for a sample *pH*-value measured using a glass type electrode system, that was calibrated with two certified *pH*-reference material solutions.

Keywords *pH*-measurement · Hydrogen electrode · *pH*-glass electrode · Harned-cell · Certified *pH*-reference materials · CRM · Uncertainty · Calibration · ISO-GUM

1 Measurement of *pH*, IUPAC 2002 Recommendation in Short

The measurement of *pH*-values is among the most widespread and frequent analytical-chemical measurement tasks. The knowledge of the hydrogen ion activity $a(H^+)$ in a water type medium expressed as a *pH*-value is important in numerous fields of activities with almost uncountable applications. To name just a few of the application fields: analytical chemistry, clinical chemistry, biochemistry, biology, water chemistry, nutrition production, environmental chemistry, geochemistry, etc. Often the *pH*-value is even of dominant importance such as e.g. in groundwater or in biological fluids. The logarithmic scale of *pH* in water ranges from 0 to 14. Values for the *pH* of a solution are mostly measured by instruments since its first definition by Søren Peter Lauritz Sørensen in 1909 [1] as the logarithm of a concentration value.

S. Wunderli (✉)
atomolmetrion GmbH, Chemin de la Binna 10, 1786 Sugiez, Switzerland
e-mail: xiongmao@gmx.ch

S. Medvedevskikh et al. (eds.), *Reference Materials in Measurement and Technology*,
https://doi.org/10.1007/978-3-030-32534-3_23

Later on it was recognised that not the concentration $c(H^+)$ but the chemical activity $a(H^+)$ is the quantity involved in Sørenson's measurements. For a long time two *pH* scales where in use and it was also recognised, that without an extra-thermodynamic assumption *pH*-values could not be accessed purely by measurement. Single-ion activities can only be evaluated if the value of the activity of a counter ion is known or defined by a suitable convention. Single ion activities are not accessible directly by measurement without such an assumption and knowledge of the activity of a counter ion. This dilemma was ultimately approached by a IUPAC working group of scientists and published as the todays valid IUPAC-definition of *pH* [2, 3], based mainly of the fundamental work of R. G. Bates, R. A. Robinson, R. H. Stokes, R. A. Durst, A. K. Covington, F. G. K. Baucke, R. Naumann and many other researchers.

Using the convention of Bates-Guggenheim [4] the activity of the chloride ion was defined. It holds for ionic strength values less than or equal 0.1 mol kg^{-1} solvent (≤0.1 m).

$$\log\{\gamma(Cl^-)\} = -\frac{A \cdot \sqrt{I}}{1 + 1.5 \cdot \sqrt{I}} \quad A \text{ Debye–Hückel constant, } I \text{ ionic strength}$$

For all temperatures and all type of solutions the ion size parameter *a* multiplied by the Debye-Hückel-*B*-constant in the above equation is set to be 1.5 $(\text{mol kg}^{-1})^{-0.5}$.

Bates-Guggenheim-Convention for the activity coefficient of chloride:

$$\log\{\gamma(Cl^-)\} = -\frac{A \cdot \sqrt{I}}{1 + B \cdot a \cdot \sqrt{I}} = -\frac{A \cdot \sqrt{I}}{1 + 1.5 \cdot \sqrt{\frac{I}{m^0}}}$$

A Debye-Hückel constant or Debye-Hückel slope:

$$A = \left(\frac{1}{4 \cdot \varepsilon_0 \cdot \varepsilon_r \cdot k \cdot T}\right)^{3/2} \cdot e^3 \cdot (2 \cdot N_A \cdot \rho_w)^{1/2} \cdot \frac{1}{\pi \cdot \ln 10}$$

A	0.5062 $(\text{mol kg}^{-1})^{-0.5}$ at 20 °C
$B.a$	$B{\cdot}a = 1.5$ $(\text{mol kg}^{-1})^{-0.5}$ *a* is the ion size parameter, set fix for 5 °C–50 °C, Debye-Hückel *B*-constant
$I = \frac{1}{2} \cdot \sum_i m_i \cdot z_i^2$	ionic strength on the molal scale (m_i: mol per kg solvent)
e	elementary charge of the electron: $1.602176634 \times 10^{-19}$ C (exact)
ε_0	electric field constant (vacuum electric permittivity): $8.8541878128(13) \times 10^{-12}$ F m^{-1}
ε_r	dielectricity constant of water: 80.2 at 20 °C [5]
k	Boltzmann constant: 1.380649×10^{-23} J K^{-1} (exact)
N_A	Avogadro constant: $6.02214076 \times 10^{23}$ mol^{-1} (exact)
ρ_w	density of water: 998.2067(8) kg m^{-3} at 20 °C [6]
T	absolute temperature K

Standard uncertainties are given in parentheses ( ) and represent the last digits of the value.

Data are from CODATA: https://pml.nist.gov/cuu/Constants/.

These temperature dependent Debye-Hückel constants *A* are numerically calculated as the following values: 0 °C: 0.4922, 5 °C: 0.4953: 10 °C: 0.4987, 15 °C: 0.5024, 20 °C: 0.5062, 25 °C: 0.5103, 30 °C: 0.5146, 35 °C: 0.5191, 40 °C: 0.5239.

In practice the IUPAC definition is useful in the *pH* range of 3–10. For seawater with higher ionic strength values up to about 0.93 mol kg^{-1} an approach based on the semi-empirical theory, the ion-interaction theory of Pitzer [7], which uses a virial-expansion-equation for the Gibbs-free energy, is more appropriate but not yet standardised on an international level. These Pitzer-type semi-empirical activity coefficients can be calculated for complex electrolyte mixtures up to high ion strengths values [8, 9].

The *pH*-value is defined as follows:

$$pH = -\log\{a_H\} = -\log\left\{\frac{\gamma_H \cdot m_H}{m^0}\right\}$$

a_H activity of the hydrogen ion H^+
γ_H molal activity coefficient of the hydrogen ion H^+
m_H molality of the hydrogen ion H^+
m^0 standard-molality: 1 mol kg^{-1} solvent (water)

This definition requires the knowledge of a single ion activity. A measurement of a single ion activity is impossible and a sound estimation or a convention for the activity of the counter ion (chloride in this case) is necessary in order to tackle the problem.

The present day method of measurement for *pH* is broadly accepted and recommended by IUPAC. It was published in the year 2002 by a working group of experts [2].

Before this time two different *pH*-scales were in use. It was clear that such a condition was untenable from a metrological point of view, although the inconsistencies were not great [10].

The recommended electrochemical cell by IUPAC used for primary standards measurements is without transference and is named Harned-cell:

$$Pt\,|\,H_2\,|\,\text{buffer S}, Cl^-\,|\,AgCl\,|Ag$$

In this redox-cell hydrogen is oxidised on the surface of a platinum electrode producing hydrated protons. A pictorial description may be as such: Molecular hydrogen H_2 as a gas is energetically favorably adsorbed on the solid surface of the platinum electrode, the single bond H–H is weakened through adsorption, possibly broken up, resulting in two adsorbed hydrogen atoms per molecule, subsequently these adsorbed hydrogen atoms are oxidised and immediately hydrated by water molecules in a concerted way producing the hydrated protons.

The two conducting electrons from the hydrogen molecules in the platinum electrode are moving through the metallic conductor from the cathode towards the anode where silver chloride is reduced to silver atoms that frees chloride ions from the solid AgCl. The chloride ion diffuses into the water based electrolyte solution.

The cell reaction is therefore shortly described as:

$$\tfrac{1}{2}\mathrm{H_2} + \mathrm{AgCl} \rightarrow \mathrm{Ag(s)} + \mathrm{H^+_{aq}} + \mathrm{Cl^-_{aq}}$$

The measured potential difference, referenced to a partial pressure of the hydrogen gas of the standard atmosphere 1 atm (101325 Pa) is expressed by the following approach in a Nernst equation:

$$E_I = E^0 - \left(\frac{R \cdot T}{F} \cdot \ln 10\right) \cdot \log\left\{\left(\frac{m_\mathrm{H} \cdot \gamma_m}{m^0}\right) \cdot \left(\frac{m_\mathrm{Cl} \cdot \gamma_\mathrm{Cl}}{m^0}\right)\right\}$$

E^0	standard potential of the Ag/AgCl reference electrode
R	molar gas constant $R = N_A \cdot k$: 8.314462618 ... J mol^{-1} K^{-1} (exact)
T	absolute temperature K
F	Faraday constant $F = N_A \cdot e$: 96485.33212 ... C mol^{-1} (exact)
γ	activity coefficient of hydrated proton rep. chloride
m_H, m_Cl	molality of hydrated proton and of the chloride
m^0	the standard molality (1 mol kg^{-1} solvent)

Data are from CODATA: https://pml.nist.gov/cuu/Constants/.

For the equation to be valid the lead attached to the hydrogen electrode has to be connected with the minus pole of the high–impedance voltmeter.

The measured potential is linearly proportional to the absolute temperature T. E^0 is the standard potential for the silver/silver chloride electrode with respect the standard hydrogen electrode, which is the zero point for all temperatures.

The so called acidity function $p(a_H \cdot \gamma_H)$ results from this equation [11]. The operator p denominate the negative decadic logarithm:

$$a_H = \frac{m_H \cdot \gamma_H}{m^0}$$

$$p(a_H \cdot \gamma_{Cl}) = -\log\{a_H \cdot \gamma_{Cl}\} = \left(\frac{E_I - E^0}{(R \cdot T/F) \cdot \ln 10}\right) + \log\left\{\frac{m_{Cl}}{m^0}\right\}$$

By measuring the potential of a Harned cell with the solution in question and e.g. three different solutions with and added sodium or potassium chloride (m_{Cl} range: 0.005 mol kg^{-1} to 0.02 mol kg^{-1}) with known ion strength and linear extrapolation of the acidity function to zero molality for chloride. This procedure gives the *pH*-value of the primary standard (PS):

$$pH\,(\mathrm{PS}) = \lim_{m_{Cl} \to 0}\left\{\frac{(E_I - E^0)}{[RT/F]}\ln 10\right\} + \log\left\{\frac{m_{Cl}}{m^0}\right\} - \frac{A\sqrt{I}}{\left[1 + 1.5 \cdot \sqrt{\frac{I}{m^0}}\right]}$$

The molal activity coefficients of the chloride ions in the three solutions mentioned above are calculated using the Bates-Guggenheim Convention (cf. above).

For the practical realisation of a hydrogen electrode guidance can be found in Chap. 5 of the handbook of reference electrodes [12].

In practice the operation of a hydrogen electrode is fundamental and used to measure values für primary and secondary *pH*-standards. But its operation is not suitable for applications in field laboratories. Therefore in most field laboratories the combined *pH*-glass electrodes with either fine pored diaphragm or a glass sleeve or membrane or a capillary type systems are in use.

By this it becomes unavoidable that the diffusion potential or junction potential between the sample solution and the electrolyte solution of the outer reference electrode system is effective and has to be considered for a correct *pH*-measurement.

On the following pages the uncertainty evaluation of *pH*-values measured by a combined *pH*-glass electrode system is outlined in detail step by step as suggested by the EURACHEM-CITAC Guide 3rd.ed. 2012 (Quantifying uncertainty in analytical measurement) or the JCGM-100:2008 guide (Guide to the Expression of Uncertainty in Measurement, which is the so called ISO-GUM with minor corrections). Both documents are available free of charge as .pdf-documents on the BIPM (JCGM-100) and the EURACHEM homepage. Additional suitable informations on potentiometric *pH*-measurement with glass electrodes are given by a number of authors in the literature references [11, 13–30]. Part of their approaches are used in the following elaboration.

2 Uncertainty Evaluation for a *pH*-Value Measured Using a Combined *pH*-Glass Electrode

Step 1: Specification of the measurand

The combined *pH*-glass electrodes contain an inner as well as an outer reference electrode system as well as a proton selective glass membrane and a diaphragm or other connections to the outer reference electrode. Special glasses have the ability to reversibly take up hydrated protons in their network. The selectivity is never perfect and leads to the so called alkali error which is minimised by using suitable glass compositions. The selective surface uptake of hydrated protons results in local charge separation, which manifests itself as a change in Galvani-potential at the boundary.

The silver/silver chloride electrodes [12, Chap. 5.2.1 by K. Maksymiuk, A. Michalska, A. Kisiel, Z. Galus] serve as connecting links between the electron-conducting part of the measuring system and the ion-conducting part. At each phase boundary the potential changes (work per charge). The potentiometric measurement device shows the sum of all potential differences (galvanic voltages, diffusion voltages, ohmic resistances) over the entire cell. Ideally, the potential difference at the interface between the proton selective glass membrane and the sample solution changes in proportion to the logarithm of the activity of the hydrated proton according to Nernst. In practice, however, the measured cell potential is also dependent on the size of the diffusion potential resulting from the different mobilities u_i of the ions between the electrolyte of the outer reference and the sample solution. Strongly acidic and strongly basic solutions generally lead to a large diffusion potential, since both the hydrated proton H^+ and the hydroxyl ion OH^- are much more mobile than other ions due to their special mechanism of charge conduction.

The schematic structure of a pH glass electrode assembly as a cell diagram shows seven phase boundaries:

Ag | AgCl | KCl 3M, buffer | glass membrane | sample solution a_H || KCl 3M | AgCl | Ag.
Ag φ_1 AgCl φ_2 KCl 3M, buffer φ_3 glass membrane φ_4 sample solution a_H φ_5 KCl 3M φ_6 AgCl φ_7 Ag.

Since only the Galvani-potential difference φ_4 is actually of interest, all other voltage differences must be kept constant for calibration and sample measurements. In addition to the potential changes at the interfaces, the ohmic resistances of the conducting phases are added. Virtually no current flows, as the internal resistance of the pH-voltmeter used is very high (usually approx. 10^{12} Ω–10^{15} Ω). At the phase boundaries of the various materials, a charge separation takes place, which leads to galvanic voltages. These cannot be determined individually since there are always at least two such phase boundaries involved in a cell. The sum of all these Galvani voltages φ_i or potential differences [12, Chap. 1, György Inzelt] occurring at each interface are measured with the potentiometer.

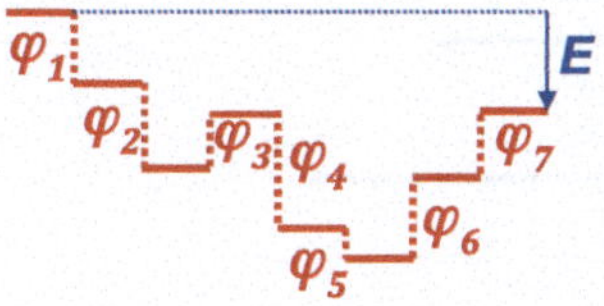

φ_1 potential difference between metallic silver and solid silver chloride

φ_2 potential difference between solid silver chloride and 3M KCl dependent on the activity a_{Cl} of the internal buffer

φ_3 potential difference between glass membrane inside and inner buffer, dependent on the pH-value of the internal buffer

φ_4 potential difference between glass membrane outside and measuring solution, dependent on the pH-value of the measuring solution

φ_5 diffusion potential (junction potential, E_J) [12, Chap. 3, Galina Tsirlina] between measuring solution and the electrolyte of the outer reference electrode, dependent on the ion mobilities $u_i = \lambda_i / F$ (λ_i: equivalence conductivity, F: Faraday constant), the concentrations c_{i} and the electrical charges z_{i} of all ions in these two phases $_{1,2}$: cf. Henderson-equation:

$$E_J = \frac{R \cdot T}{F} \cdot \frac{\left(\sum u_{i,1}^{+} \cdot c_{i,1} - \sum u_{i,1}^{-} \cdot c_{i,1}\right) - \left(\sum u_{i,2}^{+} \cdot c_{i,2} - \sum u_{i,2}^{-} \cdot c_{i,2}\right)}{\left(\sum u_{i,1}^{+} \cdot c_{i,1} \cdot z_i - \sum u_{i,1}^{-} \cdot c_{i,1} \cdot z_i\right) - \left(\sum u_{i,2}^{+} \cdot c_{i,2} \cdot z_i - \sum u_{i,2}^{-} \cdot c_{i,2} \cdot z_i\right)} \cdot \log\left\{ \frac{\left(\sum u_{i,1}^{+} \cdot c_{i,1} \cdot z_i + \sum u_{i,1}^{-} \cdot c_{i,1} \cdot z_i\right)}{\left(\sum u_{i,2}^{+} \cdot c_{i,2} \cdot z_i + \sum u_{i,2}^{-} \cdot c_{i,2} \cdot z_i\right)} \right\}$$

Henderson-equation for E_J is an approximation with approximately linear ion concentration distributions along the normal to the boundary:

Residual liquid junction potential:

$$\Delta E_{J,r} = E_{J,cal} - E_{J,sample}$$

It is the difference of the two diffusion potentials between the sample solution and the calibration solution (cf. Table for 3M KCl outer reference). The diffusion potential for the sample solution is generally unknown and can only be guessed.

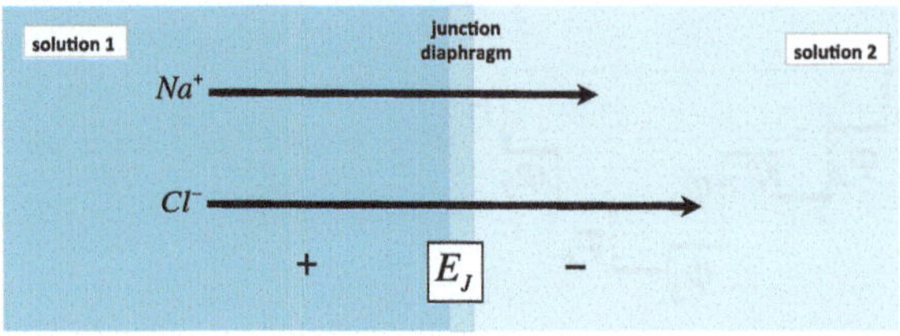

KCl 3M, outer reference

solution	E_l ; mV
NaCl, 0.001 M	- 3.96
NaCl, 0.01 M	- 2.77
NaCl, 0.1 M	- 1.23
NaCl, 1 M	1.79
NaCl, 3M	4.36

solution	E_l ; mV
$MgCl_2$, 0.001 M	- 3.58
$MgCl_2$, 0.01 M	- 2.22
$MgCl_2$, 0.1 M	-0.26
$MgCl_2$, 1 M	6.78
$MgCl_2$, 3 M	12.48

The diffusion potential is small when the electrolyte of the outer reference electrode solution contains ions of equal or similar mobility {e.g. electrolyte KCl: $u(K^+) \approx u(Cl^-)$} and when this electrolyte is highly concentrated (4.8 mol kg^{-1} KCl, counter ion logarithm large and possibly the same or similar value as the denominator, e.g. CsCl 11.3 mol kg^{-1} or RbCl 7 mol kg^{-1}, $LiSO_4$ 3M). If the diffusion potential is large, the chain voltage is often depending on the stirring speed of the sample solution. Such an influence is to be avoided.

In practice, however, the residual diffusion potential (RLJP, residual liquid junction potential) is of most importance. That's the difference between the diffusion potentials produced between the calibration and the sample solution due to mainly differences in ionic strength and ion composition. The ionic strength and the composition of calibration and sample solutions should be as similar as possible. Diffusion potentials or junction potentials can be estimated using the Henderson equation.

φ_6 Potential difference between 3M KCl and solid silver chloride ($=\varphi_2$)
φ_7 Potential difference between solid silver chloride and metallic silver ($=\varphi_1$)

For the calibration of the measuring device, there are various methodical approaches (one-point, two-point, bracketing calibration, multiple-points, etc.). The most commonly used method is the two-point calibration method with two solutions whose pH values (pH_1, pH_2) are known, i.e. certified reference materials documents. From this calibration, the practical slope ($\Delta E/\Delta pH$) value, ideally according to Nernst, would be 59.2 mV/pH at 25 °C, in general ($R \cdot T/F \cdot \ln 10$) and the zero point are obtained. The zero point of a cell voltage corresponds to the pH value at which the measured cell voltage is zero. It is usually close to $pH = 7$. The zero point is dependent on all individual voltage differences at each phase boundary of a pH electrode assembly and can change therefore.

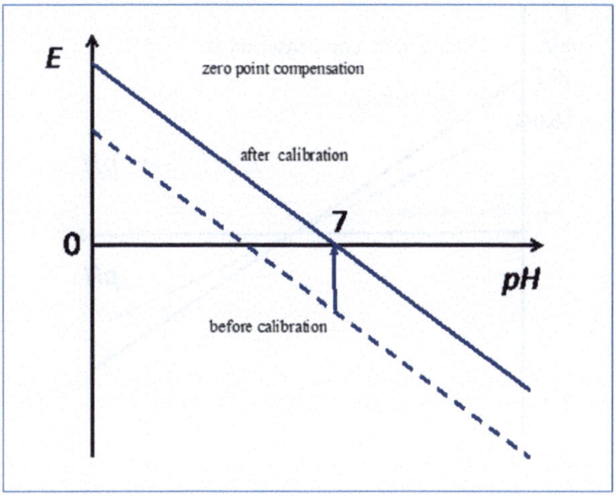

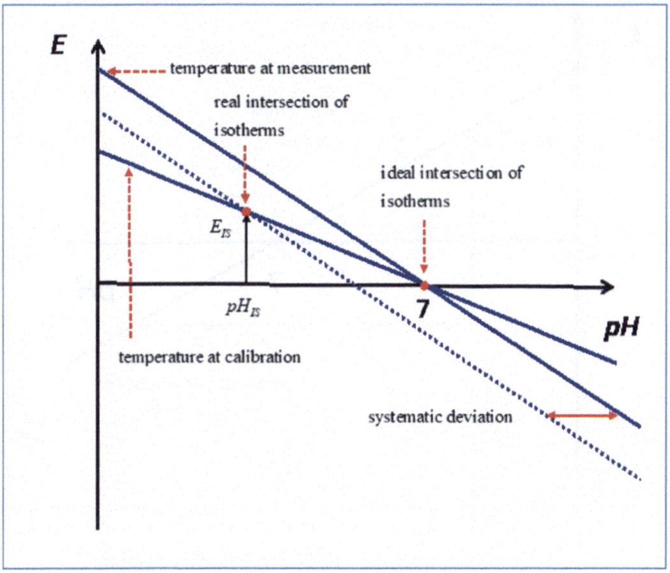

Isotherm intersection at $\{pH_{\mathrm{IS}}, E_{\mathrm{IS}}\}$ not at $\{pH = 7, E = 0\}$

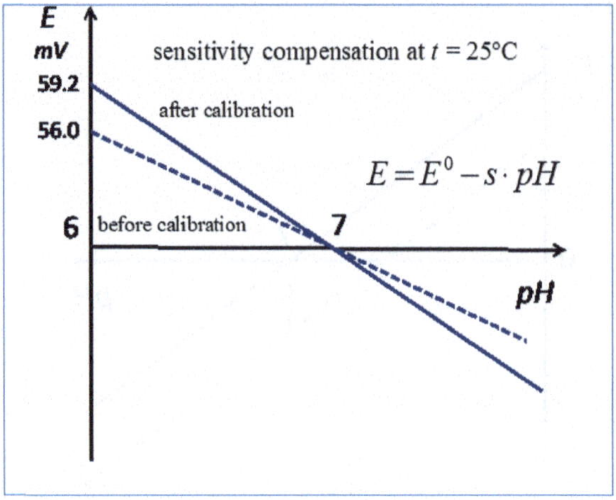

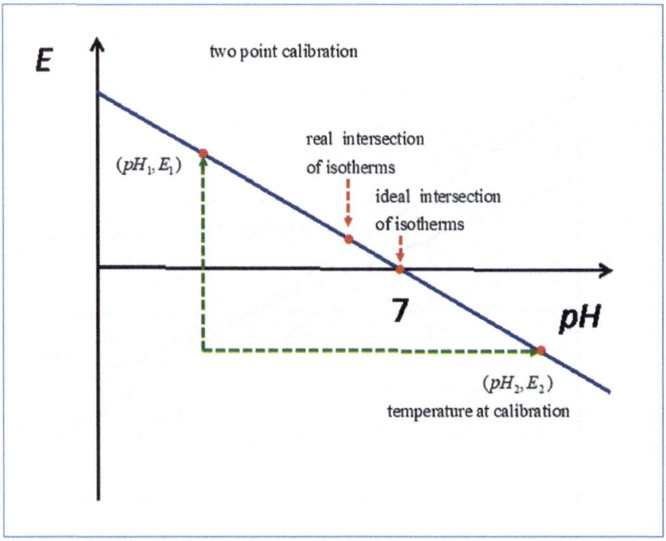

Two point calibration using CRMs at a defined temperature

The practical zero point of the cell is the intersection of the calibration line with the pH axis at $E = 0$. The calculation requires the values for the cell voltage at two different pH values:

$$pH_{0,t_1} = \frac{pH_2 \cdot E_1 - pH_1 \cdot E_2}{E_1 - E_2} \quad pH_{0,t_2} = \frac{pH_4 \cdot E_3 - pH_3 \cdot E_4}{E_3 - E_4}$$

An equilibrium voltage is read if the potential change has reached less than 0.1 mV per 10 s measurement time.

The sensitivity (slope) of the calibration line (E vs. pH) changes with temperature because the Nernst voltage is temperature dependent. The change is linear. Calibration lines recorded at different temperatures intersect at the isothermal intersection point (pH_{IS}, E_{IS}). With symmetrical cells (identical or equal inner and outer reference systems) the straight lines intersect at one point. If one uses an automatic temperature compensation with an asymmetrical cell, a systematic deviation results. The isotherm intersection point is determined using four standards at two temperatures.

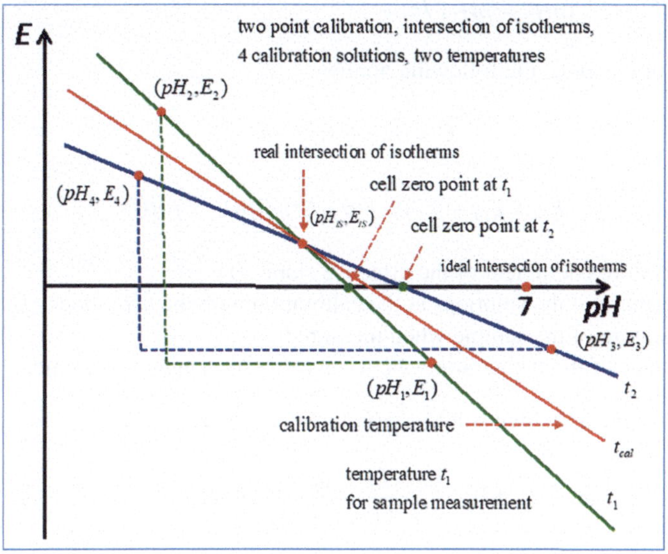

Four point calibration with CRMs at two temperatures defines the isotherm intersection point $\{pH_{IS}, E_{IS}\}$

The behaviour of the pH glass electrode can be described in the range from approx. $pH = 2$ to $pH = 12$ using a linear equation which is characteristic for a straight line bundle that passes through the fixed point (pH_{IS}, E_{IS}). The straight line (E_x vs. pH_x) is defined by the two calibration points (pH_1, E_1), (pH_2, E_2).

$$E_x = -s(t) \cdot pH_x + C \quad E_{IS} = -s(t) \cdot pH_{IS} + C \Rightarrow C = E_{IS} + s(t) \cdot pH_{IS}$$

For the experimental slope (sensitivity) at the temperature used for the calibration the following holds:

$$s = \frac{E_2 - E_1}{pH_1 - pH_2}$$

This slope can also be determined from the isotherm intercept (pH_{IS}, E_{IS}) and a calibration point (e.g., (pH_1, E_1):

$$s = \frac{(E_1 - E_{IS})}{(pH_{IS} - pH_1)} \quad \Rightarrow \quad pH_{IS} = pH_1 + \frac{(E_1 - E_{IS})}{s}$$
$$= pH_1 + \frac{(E_1 - E_{IS})}{(E_1 - E_1)} \cdot (pH_2 - pH_1)$$

The effective experimental slope $s(t)$ is temperature dependent:

$$s(t) = \left(\frac{(E_2 - E_1)}{(pH_2 - pH_1)}\right) \cdot (1 + \alpha \cdot (t - t_{cal})) = s \cdot (1 + \alpha \cdot \Delta t)$$

For the intercept C, the following applies:

$$C = E_{IS} + s(t) \cdot pH_{IS}$$

$$E_x = E_{IS} - s \cdot (1 + \alpha \cdot \Delta t) \cdot (pH_x - pH_{IS})$$

α temperature coefficient of the effective slope $s(t)$
t_{cal} temperature of the solutions at the calibration with the pH-values pH_1, pH_2
t temperature of the sample when measured
Δt difference of the temperatures between calibration and measurement $(t - t_{cal})$

The pH of the unknown solution pH_x turns out to be:

$$pH_x = \frac{E_{IS} - E_x}{s \cdot (1 + \alpha \cdot \Delta t)} + pH_{IS}$$

Insertion of the experimentally determined slope s:

$$pH_x = \frac{(E_{IS} - E_x) \cdot (pH_1 - pH_2)}{(E_2 - E_1) \cdot (1 + \alpha \cdot \Delta t)} + pH_{IS}$$

$$\boxed{pH_x = \frac{(E_{IS}-E_x)\cdot(pH_1-pH_2)}{(E_2-E_1)\cdot(1+\alpha\cdot\Delta t)} + pH_1 + \frac{(E_1-E_{IS})}{(E_2-E_1)} \cdot (pH_1 - pH_2)}$$

This is the equation for the measurand pH_x, which is the pH-value attributed to the sample.

Certain influences, such as the inflow, the ionic strength I, the carbon dioxide CO_2 content of the air, chemical effects at large and small pH values on the integrity of the glass membrane are not taken into account here.

Step 2: Identification of the sources of uncertainty

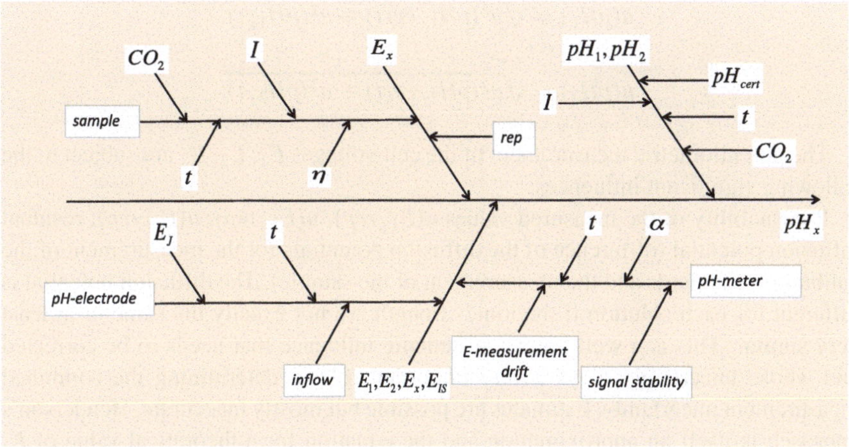

The fish-bone diagram for the pH-measurement helps to identify influence parameters, but it gives no information about the functional relationship of the variables that are part of the model equation for the measurand.

List of main influences:

pH_x	measurand, *pH* value of the sample solution (measurement solution)
pH_1, pH_2	*pH*-values of the calibration solutions (certified values for pH_1, pH_2)
E_1, E_2	measured voltages of the cell for the calibration solutions
E_x	measured voltage of the cell for the sample solution
E_J	diffusion potential (junction potential E_5 at diaphragm)
I	ionic strength ($I = \frac{1}{2} \cdot \sum_i c_i \cdot z_i^2$, ion i, concentration c_i, ion charge z_i)
η	viscosity of the sample solution
Rep	repeatability
CO_2	influence of the carbon dioxide CO_2 contained in air (may be relevant for basic sample and calibration solutions if no precautions are taken)
Effect from inflow	sensitivity from stirring
t	temperature influences slope, intersection point of isotherms, etc.
α	temperature coefficient of effective Nernst slope $s(t)$

Uncertainties arising from the *pH* values of the standards used pH_1, pH_2 and uncertainties of the measured voltage values E_1, E_2, E_x, E_{IS} are to be distinguished.

For the *pH* standards used, both the value from the certificate and the effect of temperature on these values must be taken into account. The respective contributions should be additive. The standard uncertainty of the pH values for the standards can therefore be stated as follows:

$$u(pH_1) = \sqrt{u^2(pH_1, cert) + u^2(pH_1, t)}$$

$$u(pH_2) = \sqrt{u^2(pH_2, cert) + u^2(pH_2, t)}$$

The potentiometric measurement of the cell voltages E_1, E_2, E_x are subject to the following significant influences:

Repeatability of the measured values $u(E_1, rep)$, $u(E_2, rep)$, $u(E_x, rep)$, residual diffusion potential (difference of the diffusion potentials for the measurement of the calibration standards and the measurement of the sample). The diffusion potential is different for each solution if the ionic strength are not exactly the same or at least very similar. This is a well-known systematic influence that needs to be corrected and whose uncertainty plays a very important role in determining the combined measurement uncertainty. Estimates are possible but mostly inaccurate. Henderson's approach is itself an approximation and the equation for a theoretical value of E_J can only be used if the composition as well as the geometry of both liquid phases are known. For field laboratories such considerations usually go too far. Although wrong, this systematic effect is only calculated with an uncertainty for the measured values of the calibration standards E_1 and E_2 and not additionally as bias ΔpH_{EJ}: $u(E_1, \Delta E_{J,1})$, $u(E_2, \Delta E_{J,2})$:

For the sample solution, this systematic influence is approximately taken into account. There remains an instrumental drift: $u(E_x, drift)$.

A possible stirring effect is already included in the consideration of the diffusion potential. The above effects are all additive.

The uncertainty of the measured potentials leads to the following approach:

$$u(E_1) = \sqrt{u^2(E_1, rep) + u^2(E_1, \Delta E_{J,1})}$$

$$u(E_2) = \sqrt{u^2(E_2, rep) + u^2(E_2, \Delta E_{J,2})}$$

$$u(E_x) = \sqrt{u^2(E_x, rep) + u^2(E_x, drift)}$$

The parameters E_{IS}, α, Δt have no influence factors of second order. The corresponding uncertainties are therefore simple: $u(E_{IS})$, $u(\alpha)$, α, $u(\Delta t)$.

When measuring the unknown sample, at least the following four influencing parameters can be effective, some of which have already been taken into account.

1. Repeatability of the measurement (*rep*), 2. Uncertainty from the limited display of the pH meter (*read*), 3. Uncertainty due to drift of the measuring system (*drift*), 4. Temperature effect which affects the sensitivity (*slope*). This last influence has not yet been taken into account in the standards for calibration. The influences 1 and 3 are already taken into account (see $u(E_x)$ above). The influence of the deviating

temperature is taken into account in $u(\Delta t)$. Therefore it would be useful to extend the model equation for the measurand by adding an influence term of the *pH* measurement of the sample δpH_{xm}. The expectation value of which is zero, but not its uncertainty stemming from repeatability, readability and drift:

A more advanced model for the measurand pH_x is:

$$\boxed{pH_x = \frac{(E_{IS}-E_x)\cdot(pH_1-pH_2)}{(E_2-E_1)\cdot(1+\alpha\cdot\Delta t)} + pH_1 + \frac{(E_1-E_{IS})}{(E_2-E_1)}\cdot(pH_1-pH_2) + \delta pH_{xm}}$$

$$u(\delta pH_{xm}) = \sqrt{u^2(\delta pH_{xm}, rep) + u^2(\delta pH_{xm}, read) + u^2(\delta pH_{xm}, drift)}$$

As repeatability and drift are now included in $u(\delta pH_{xm})$, the standard uncertainty becomes $u(E_x) = 0$.

Step 3: Quantification of the uncertainties and conversion into standard uncertainties

Information about the pH-meter used:

pH-meter 780 from Metrohm. The pH meter needs the value $E_{IS} = 0$ mV for the isothermal intersection. The resolution is 0.1 mV in the mode for the potential measurement and 0.001 for the *pH* display resolution. The uncertainty of the temperature measurement is 0.1 °C. There are no data available for drift.

Information about the pH-electrode sensor system used:

pH-glass electrode filled with Ag/AgCl with 3M KCl inner and outer reference systems, ceramic diaphragm, integrated Pt1000 thermocouple, electrode zero point at 0 ± 15 mV ($pH = 7$), isotherm intersection E_{IS} at 0 ± 15 mV. Electrode slope larger than 0.97 (relative to Nernst). Temperature range for application 0–80 °C.

In the literature, values for the residual diffusion potential for dilute buffer solutions of about 0.1–1.6 mV can be found. This corresponds to 0.002–0.027 *pH* units at 25 °C, or $([1 - \exp\{-\Delta E\cdot F/(R\cdot T\cdot \ln(10)\}]$.

Theoretically:

$$\alpha_{(25)} = \frac{1}{\Delta t}\left(\frac{R\cdot 298.15/F\cdot \ln\{10\}}{R\cdot T/F\cdot \ln\{10\}} - 1\right) = \frac{1}{(t-25)}\cdot\left(\frac{298.15}{273.15+t} - 1\right)$$
$$= 3.354\times 10^{-3}\,\mathrm{K}^{-1}$$

Experimentally:

$$\alpha_{(25)} = \frac{1}{(t_2-t_1)}\left(\frac{s(t_1)}{s(t_2)} - 1\right) = \frac{1}{(t_2-t_1)}\cdot\left(\frac{(E_3-E_4)\cdot(pH_2-pH_1)}{(pH_4-pH_3)\cdot(E_1-E_2)} - 1\right)$$

Information about the certified pH-calibration solutions:

Three certified pH standard reference solutions with traceable pH values at 25 °C:

$$pH = 4.00 \pm 0.02, pH = 7.00 \pm 0.02, pH = 10.00 \pm 0.02.$$

The temperature dependencies of the three references are in order 0.001, 0.002, 0.01 pH units per °C (indication of the manufacturer, certificate).

It is somewhat unfortunate, that most commercial suppliers of calibration solutions do not offer any information about the ionic strength of the certified pH-reference material solutions. In order to minimise the residual junction potential such information would be very usefull.

Uncertainty components (measured: Type A, estimated: Type B):

$u(E_{IS})$	1 mV	B
$u(\alpha)$	0.002×10^{-3} K^{-1}	B
$u(\Delta t)$	1 °C	A
$u(E_x)$	0 mV	in $u(\delta pH_x)$
$u(E_x) = \sqrt{u^2(E_x, rep) + u^2(E_x, drift)}$		
$u(E_x, rep)$	0 mV	in $u(\delta pH_x)$
$u(E_x, drift)$	0 mV	in $u(\delta pH_x)$
$u(E_1)$	1.56 mV	A
$u(E_1) = \sqrt{u^2(E_1, rep) + u^2(E_1, \Delta E_{J,1})}$		
$u(E_1, rep)$	0.5 mV	A
$u(E_1, \Delta E_{J,1})$	1.48 mV	B
$u(E_2)$	1.56 mV	A
$u(E_2) = \sqrt{u^2(E_2, rep) + u^2(E_2, \Delta E_{J,2})}$		
$u(E_2, rep)$	0.5 mV	A
$u(E_2, \Delta E_{J,2})$	1.48 mV	B
$u(pH_1)$	0.02	B
$u(pH_1) = \sqrt{u^2(pH_1, cert) + u^2(pH_1, t)}$		
$u(pH_1, cert)$	0.02	B
$u(pH_1, t)$	0.001	B
$u(pH_2)$	0.0201	B
$u(pH_2) = \sqrt{u^2(pH_2, cert) + u^2(pH_2, t)}$		
$u(pH_2, cert)$	0.02	B
$u(pH_2, t)$	0.002	B
$u(\delta pH_{xm})$	0.0201	B

(continued)

(continued)

$u(\delta pH_{xm}) = \sqrt{u^2(\delta pH_{xm}, rep) + u^2(\delta pH_{xm}, read) + u^2(\delta pH_{xm}, drift)}$		
$u(\delta pH_{xm}, rep)$	0.002	A
$u(\delta pH_{xm}, read)$	0.001	B
$u(\delta pH_{xm}, drift)$	0.02	B

Step 4: Propagation of the standard uncertainties for the calculation of the combined measurement uncertainty

The combined measurement uncertainty is obtained by differentiation of the measurand according to all known variables (second order contributions are not considered):

$$u_c(pH_x) = \sqrt{\begin{array}{l}\left(\frac{\partial pH_x}{\partial E_1}\right)^2 \cdot u^2(E_1) + \left(\frac{\partial pH_x}{\partial E_2}\right)^2 \cdot u^2(E_2) + \left(\frac{\partial pH_x}{\partial E_{IS}}\right)^2 \cdot u^2(E_{IS}) \\ +\left(\frac{\partial pH_x}{\partial E_x}\right)^2 \cdot u^2(E_x) + \left(\frac{\partial pH_x}{\partial \delta pH_{xm}}\right)^2 \cdot u^2(\delta pH_{xm}) + \left(\frac{\partial pH_x}{\partial pH_1}\right)^2 \cdot u^2(pH_1) \\ +\left(\frac{\partial pH_x}{\partial pH_2}\right)^2 \cdot u^2(pH_2) + \left(\frac{\partial pH_x}{\partial \alpha}\right)^2 \cdot u^2(\alpha) + \left(\frac{\partial pH_x}{\partial \Delta t}\right)^2 \cdot u^2(\Delta t)\end{array}}$$

The calculations of the partial derivatives are carried out according to known rules of differentiation or can be calculated approximately by means of difference quotients.

$$\left(\frac{\partial pH_x}{\partial E_1}\right) = \frac{(pH_1 - pH_2) \cdot (E_{IS} - E_x)}{(E_2 - E_1)^2 \cdot (1 + \alpha \cdot \Delta t)} + \frac{(-E_{IS} + E_1) \cdot (pH_1 - pH_2)}{(E_2 - E_1)^2} + \frac{(pH_1 - pH_2)}{(E_2 - E_1)}$$

$$\left(\frac{\partial pH_x}{\partial E_2}\right) = -\frac{(E_{IS} - E_x) \cdot (pH_1 - pH_2)}{(E_2 - E_1)^2 \cdot (1 + \alpha \cdot \Delta t)} - \frac{(-E_{IS} + E_1) \cdot (pH_1 - pH_2)}{(E_2 - E_1)^2}$$

$$\left(\frac{\partial pH_x}{\partial E_{IS}}\right) = \frac{(pH_1 - pH_2)}{(E_2 - E_1) \cdot (1 + \alpha \cdot \Delta t)} - \frac{(pH_1 - pH_2)}{(E_2 - E_1)}$$

$$\left(\frac{\partial pH_x}{\partial E_x}\right) = -\frac{(pH_1 - pH_2)}{(E_2 - E_1) \cdot (1 + \alpha \cdot \Delta t)}$$

$$\left(\frac{\partial pH_x}{\partial pH_1}\right) = \frac{(E_{IS} - E_x)}{(E_2 - E_1) \cdot (1 + \alpha \cdot \Delta t)} + \frac{(-E_{IS} + E_1)}{(E_2 - E_1)} + 1$$

$$\left(\frac{\partial pH_x}{\partial pH_2}\right) = \frac{(E_{IS} - E_x)}{(E_2 - E_1) \cdot (1 + \alpha \cdot \Delta t)} - \frac{(-E_{IS} + E_1)}{(E_2 - E_1)}$$

$$\left(\frac{\partial pH_x}{\partial \delta pH_{xm}}\right) = 1$$

$$\left(\frac{\partial pH_x}{\partial \alpha}\right) = -\frac{(pH_1 - pH_2) \cdot (E_{IS} - E_x) \cdot \Delta t}{(E_2 - E_1) \cdot (1 + \alpha \cdot \Delta t)^2}$$

$$\left(\frac{\partial pH_x}{\partial \Delta t}\right) = -\frac{(pH_1 - pH_2) \cdot (E_{IS} - E_x) \cdot \alpha}{(E_2 - E_1) \cdot (1 + \alpha \cdot \Delta t)^2}$$

Expectation values for the variables:

Influence parameter	Value	Standard uncertainty	Sensitivity factor
E_{IS}	1 mV	$u(E_{IS}) = 1$ mV	$\left(\frac{\partial pH_x}{\partial E_{IS}}\right) = 1.44 \times 10^{-4}$
α	3.334×10^{-3} K^{-1}	$u(\alpha) = 2 \times 10^{-6}$ K^{-1}	$\left(\frac{\partial pH_x}{\partial \alpha}\right) = 4.36 \times 10^{-2}$
Δt	−2.5 °C	$u(\Delta t) = 1$ °C	$\left(\frac{\partial pH_x}{\partial \Delta t}\right) = 9.31 \times 10^{-3}$
E_x	161.4 mV	$u(E_x) = 0$ mV	$\left(\frac{\partial pH_x}{\partial E_x}\right) = -1.73 \times 10^{-2}$
E_1	−2.6 mV	$u(E_1) = 1.56$ mV	$\left(\frac{\partial pH_x}{\partial E_1}\right) = 9.54 \times 10^{-4}$
E_2	172.5 mV	$u(E_2) = 1.56$ mV	$\left(\frac{\partial pH_x}{\partial E_2}\right) = 1.62 \times 10^{-2}$
pH_1	7.00	$u(pH_1) = 0.02$	$\left(\frac{\partial pH_x}{\partial pH_1}\right) = 1.02$
pH_2	4.00	$u(pH_2) = 0.0201$	$\left(\frac{\partial pH_x}{\partial pH_2}\right) = 1.74 \times 10^{-2}$
δpH_{xm}	0	$u(\delta pH_{xm}) = 0.0201$	$\left(\frac{\partial pH_x}{\partial pH_{xm}}\right) = 1$

Budget of the influences on the combined measurement uncertainty u_c graphically:

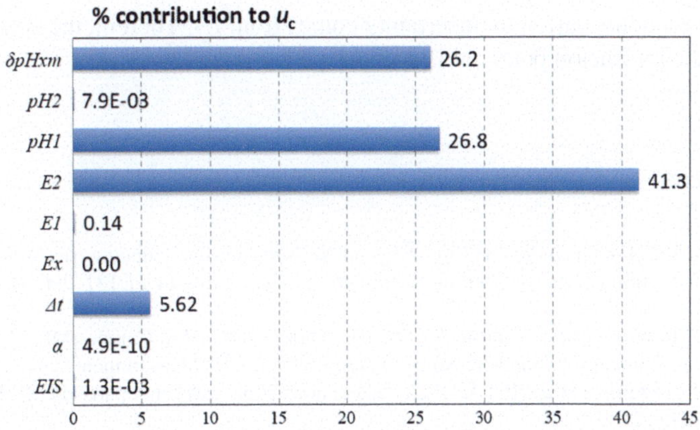

Step 5: Calculation of the expanded measurement uncertainty and reporting of the result

	Values *pH*, mV, °C^{-1}, °C	Standard-uncertainty u	Sensitivity factors	(Sensitivity factors)^2 $*u^2$	Contributions %
pH_1	7.00	0.02	1.02	4.14E−4	**26.8**
pH_2	4.00	0.0201	0.0174	1.22E−7	**7.9E−3**
δpH_{xm}	0	0.0201	1	4.04E−4	**26.2**
E_x	161.4	0	−1.75E−02	0	**0**
E_1	−2.6	1.56	9.54E−04	2.22E−6	**0.14**
E_2	172.5	1.56	1.618E−2	6.37E−4	**41.3**
E_{IS}	1	1	1.46E−01	2.07E−08	**1.3E−3**
a	3.334E−03	2.00E−06	−7.07	7.6E−15	**4.9E−10**
Δt	−2.5	1	9.43E−03	8.68E−05	**5.62**

Combined measurement uncertainty u_c:

$$u_c(pH_x) = 0.0393$$

Expanded uncertainty $U = k \cdot u_c$:

$$U(pH_x) = 2 \cdot u_c(pH_x) = 0.078(k = 2)$$

Result:

$$pH_x = \mathbf{4.167}, U(pH_x) = \mathbf{2} \cdot u_c(pH_x) = \mathbf{0.078}(k = \mathbf{2})$$

The largest contribution to uncertainty comes from E_2. Therein, the contribution of the residual diffusion potential is decisive.

References

1. Sørensen SPL, Enzymstudien II (1909) Über die Messung und die Bedeutung der Wasserstoffionenkonzentration bei enzymatischen Prozessen. Biochem Zeitschr 21:131–200, continuation 201–304
2. Buck RP, Rondinini S, Covington AK, Baucke FGK, Brett CMA, Camões MF, Milton MJT, Mussini T, Naumann R, Pratt KW, Spitzer P, Wilson GS (2001) Measurement of pH. Definition, standards, and procedures (IUPAC recommendations 2002). Working party on pH. Pure Appl Chem 74(11):2169–2200
3. Spitzer P, Pratt KW (2011) The history and development of a rigorous metrological basis for pH measurements. J Solid State Electrochem 15:69–76
4. Bates RG, Guggenheim EA (1960) Report on the standardization of *pH* and related terminology. Pure Appl Chem 1:163–168
5. Archer DG, Wang P (1990) The dielectric constant of water and Debye-Hückel limiting law slopes. J Phys Chem Ref Data 19(2):371
6. Tanaka M, Girard G, Davis R, Peuto A, Bignell N (2001) Recommended table for the density of water between 0 °C and 40 °C based on recent experimental reports. Metrologia 38:301–309
7. Pitzer KS (1991) Activity coefficients in electrolyte solutions, 2nd edn. CRC Press, Boca-Raton. ISBN 0849354153
8. Berdat D, Andres H, Wunderli S (2009) Development of suitable ISE measurement procedures for SI-traceable chemical activity determination. Chimia 63(10):671–677
9. Bastkowski F, Spitzer P, Eberhardt R, Adel B, Wunderli S, Berdat D, Andres H, Brunschwig O, Máriássy M, Fehér R, Demuth C, Gonzaga FB, Borges PP, da Silva Junior WB, Vospělová A, Vičarová M, Srithongtim S (2013) Pitzer ion activities in mixed electrolytes for calibration of ion-selective electrodes used in clinical chemistry. Accred Qual Assur 18:469–479
10. Covington AK, Bates RG, Durst RA (1985) Definition of *pH* scales, standard reference values, measurement of *pH* and related terminolohgy. Pure Appl Chem 57(3):531–542
11. Bates RG (1973) Determination of pH. Theory and practice, 2nd edn. Wiley, Hoboken. ISBN 0-471-05657-2
12. Spitzer P, Wunderli S (2013) Handbook of reference electrodes. In: Inzelt G, Lewenstam A, Scholz F (eds) Chapter 5.1: Reference electrodes for aqueous solutions. The hydrogen electrode. Springer, Berlin
13. Leito I, Strauss L, Koort E, Pihl V (2002) Estimation of uncertainty of routine pH measurement. Accred Qual Assur 7:242–249
14. Leito I, Kadis R (2010) Evaluation of residual liquid junction potential contribution to the uncertainty in pH measurement: a case study on low ionic strength natural waters. Anal Chim Acta 664(2):129–135
15. Tauber G (2010) Industrielle pH-Messung – Beiträge der Temperatur zur Messunsicherheit. Tech Mess 77(3):150–155
16. Tauber G (2009) Industrielle pH-Messung – Beiträge der Diffusionspotenziale zur Messunsicherheit. Tech Mess 76(6):308–316
17. Wilhelm T, Pechstein T (2010) Einflüsse realer Einsatzbedingungen auf die pH-Messung mit Glaselektroden. Tech Mess 77(3):140–149

18. Naumann R, Alexander-Weber Ch, Eberhardt R, Giera J, Spitzer P (2002) Traceability of pH measurements by glass electrode cells: performance characteristics of pH electrodes by multi-point calibration. Anal Bioanal Chem 374:778–786
19. pH-Messung—pH-Messung von wässerigen Lösungen mit pH-Glaselektroden und Abschätzung der Messunsicherheit, DIN 19268 (2007)
20. Water quality—Determination of pH. ISO 10523 (2008) 2nd edn
21. Degner R (2009) pH-Messung, Der Leitfaden für Praktiker. Wiley-VCH Verlag GmbH & Co. KGaA, Weinheim. ISBN 978-3-527-32359-3
22. Galster H (1990) pH-Messung, Grundlagen, Methoden, Anwendungen, Geräte, VCH-Verlagsgesellschaft. ISBN 3-527-27836-2
23. Spitzer P, Eberhardt R, Schmidt I, Sudmeier U (1996) Improved traceability of pH measurements. Fresenius J Anal Chem 356:178–181
24. Meinrath G, Spitzer P (2000) Uncertainties in determination of pH. Mikrochim Acta 135:155–168
25. Schwabe K (1976) pH-Messtechnik, 4th edn. Verlag Theodor Steinkopff, Dresden
26. Schwabe K (1980) pH-Messung. Akademie-Verlag, Berlin
27. Spitzer P (2001) Traceable measurements of pH. Accredit Qual Assur 6:55–60
28. Spitzer P, Adel B, Werner B, Jehnert D (2010) Zuverlässige Kalibrierung von pH-Messeinrichtungen. Tech Mess 77(3):156–161
29. Meinrath G, Spitzer P (2001) Importance of traceable pH measurements in science and technology. In: 166th PTB-Seminar, PTB-ThEx-25, Physikalisch-Technische Bundesanstalt, Braunschweig u. Berlin. ISBN 3-89701-903-5
30. Bates RG, Popovych O (1981) The modern meaning of pH. CRC Crit Rev Anal Chem 10(3):247–278

The manufacturer's authorised representative in the EU is Springer Nature Customer Service Centre GmbH, Europaplatz 3, 69115 Heidelberg, Germany. If you have any concerns regarding our products, please contact ProductSafety@springernature.com

Printed and bound by CPI Group (UK) Ltd, Croydon, CR0 4YY
30/09/2026
02225462-0007